CHEMISTRY
FOR
BREAKFAST

DR. MAI THI NGUYEN-KIM

CHEMISTRY
FOR
BREAKFAST

The Amazing Science of Everyday Life

Translated by
SARAH PYBUS

Illustrations by
CLAIRE LENKOVA

GREYSTONE BOOKS
Vancouver/Berkeley

Greystone Books Ltd.
greystonebooks.com

Cataloguing data available from Library and Archives Canada
ISBN 978-1-77164-748-9 (pbk)
ISBN 978-1-77164-749-6 (epub)

Copy editing by Rowena Rae
Proofreading by Meg Yamamoto
Cover and text design by Belle Wuthrich
Illustrations by claire Lenkova

Printed and bound in Canada on ancient-forest-friendly paper by Friesens

Greystone Books gratefully acknowledges the Musqueam, Squamish, and
Tsleil-Waututh peoples on whose land our office is located.

Greystone Books thanks the Canada Council for the Arts, the British
Columbia Arts Council, the Province of British Columbia through the Book
Publishing Tax Credit, and the Government of Canada for supporting our
publishing activities.

The translation of this work was supported by a grant from the Goethe-Institut.

For my mom

My mom and dad are the most loving parents in the world. Such super-latives are rare for me, but I can use them with a clear conscience when it comes to my parents. They are a true team; they have fought together and have always put their own interests aside; they made a new home in what was then a foreign country so that my brother and I could lead the privileged lives we enjoy to this day.

Often I talk only about my dad; not only is he a wonderful father, hus-band, and chemist, he also inspired me and my brother to take up chemistry ourselves. But here, now, I would like to dedicate this book to my mom. She has been my greatest influence. She is the one who decided to stay home and commit herself to loving and caring for me and my brother. She is the one who has hugged me, supported me, and motivated me every single day. Her boundless devotion has made me the person I am today. This book would never have happened without my mom—so if you like it, she's the one you should thank.

Contents

Foreword

I WAS A pretty ugly baby. I was born with jaundice and refused to eat or drink. Seriously concerned, my parents did everything they could to feed me as much as possible—and continued to do so long after my health improved. As a result, I became a real butterball. When my hair began to grow, I developed the asymmetrical receding hairline of an elderly man. Naturally, my parents thought I was the most beautiful baby in the world.

As a chemist, I sometimes feel like the mother of an ugly child whose beauty only I can see. Most people see chemistry as evil, poisonous, artificial. Or they remember how much they hated it at school and couldn't wait to drop the class. Convincing these people that my baby is beautiful is a science in itself.

At best, people have no idea what chemistry entails. They look at me, wide-eyed and clueless, and ask, "And what can you do with chemistry?"

Sometimes I'd like to shake them by the shoulders and yell "EVERYTHING!!! Chemistry is EVERYTHING!!!" For example, delicious food is one of my earliest associations with chemistry—my father is both a chemist and an excellent cook. He told me that all chemists are good cooks. If you can't cook, then you're not a good chemist. When I started to develop an interest in cosmetics at the age of thirteen, my father was able to explain everything involved there too— the chemical structure of color pigments, how volumizing hair spray works, and the pH value of face cream. For me, chemistry has always been part of everyday life.

Since studying chemistry, I have been beyond help. Whether I'm drinking coffee, brushing my teeth, or exercising, I think about adenosine receptors, fluorides, and metabolic enzymes. If I go for a walk in the sunshine, I think about melanin and vitamin D; if I cook noodles, my mind turns to boiling-point elevation and starch polymers. And I've become a pretty good cook too—I wouldn't be a good chemist otherwise.

Many people have very specific ideas not only about chemistry itself, but also about the people who work in the field. I'm often told that I don't look like a chemist. The success of The Big Bang Theory may have made it socially acceptable to be a nerd, but it also showcased many clichés— for example, that it's categorically impossible to be an expert and have social skills. This is just one of the many clichés we scientists have to fight. Scientists are unfamiliar creatures who spend their lives in laboratories or surrounded by bookshelves. Nobody knows what we look like, whether we have hobbies, or whether we even have friends. Are scientists humans too? There's just no way of knowing.

During my doctorate, I decided to start a YouTube channel called *The Secret Life of Scientists*. I wanted my videos to give a face to science. I didn't just want to show how cool science is, but also how cool scientists are. This mission is like a complex research project, and I'm still working on it today. I now also produce the *maiLab* YouTube channel for the German content network *funk* and present the German TV series *Quarks*.

So why write a book as well? Because I want to really let off some steam. This book invites you into my chemist's brain and provides a brief insight into my everyday life as a science journalist and YouTuber. Above all, however, I want you to read this book, look deep into the eyes of chemistry, and succumb to its irresistible charm. And if I'm right to believe that humans are curious creatures, then reading this book will show you not only that chemistry really is everything, but maybe even how beautiful this science can be.

Obsessed with Chemistry

BRRING-BRRING-BRRING!!!

I almost fall out of bed in fright. My heart is racing.

Furious, I want to scream "Matthiiiiiiiaaaaaaaaaaas," but my linguistic faculties don't quite seem to be working yet. My body's in a strange sort of limbo, halfway between dozing and hand-to-hand combat. I throw myself at Matthias/his cell phone and flail around until I manage to turn off his awful alarm. It's six in the morning, dammit!

Matthias has a terrible habit of getting up in the middle of the night to go running, at least twice a week—well, I consider 6 a.m. to be the middle of the night. Unfortunately, this always means that I need to wake up slightly before him so that my day doesn't start with an influx of stress hormones.

I prefer to be woken by a barely audible tinkling, as if by a fairy—otherwise I start my day with palpitations—but Matthias needs at least 100 decibels and this awful BRRING-BRRING-BRRING to wake up at all. So I usually set my alarm for one minute before his to mentally prepare myself for the stress. Last night when I set my alarm, however, I didn't know he was planning to exercise so early this morning.

I open the drapes to reduce Matthias's melatonin level.

"Matthias," I eventually manage to say.

"Hmm," he mumbles, still half asleep. Unbelievable.

THE MOLECULE *MELATONIN* is also popularly known as the "sleep hormone." It's produced in the pineal gland, a small gland in the middle of the brain. There's a very good reason for its nickname; melatonin plays an important role in our circadian rhythm (from the Latin *circa dies*, "around the day"), our internal sleep-wake cycle. The higher our melatonin level, the more tired we feel. Conveniently, light helps to reduce melatonin concentration, and it finally seems to be having an effect on Matthias.

Melatonin

I feel compelled to see the world in molecules, but it's a nice feeling. Basically, I'm obsessed with chemistry. It makes me sad to think about all the nonchemists going about their lives, not thinking about molecules at all. They don't even know what they're missing. Ultimately, anything that interests you as an individual can somehow be explained through chemistry. And you, dear reader, are in fact nothing more than a mound of molecules reading about molecules. And chemists are mounds of molecules thinking about molecules. It's almost spiritual.

So what does my morning look like in molecules?

How well we wake up in the morning is largely determined by two molecules. We need less of one—melatonin—and more of the other, the stress hormone *cortisol*, which is automatically released in the morning. "Stress hormone" might not sound great, but a moderate amount of cortisol simply helps us to get going. Normally, this wonderful service provided by our body doesn't even need an alarm clock. The BRRING-BRRING-BRRING may have been a bit too much and has triggered an actual fight-or-flight response in my body, an ingenious emergency system tried and tested since primeval times to save us from mortal danger.

Like pain, stress is generally a welcome bodily response. Pain tells us that something isn't right, and stress helps to save our lives. Imagine that you're walking around in the Stone Age and a saber-tooth tiger crosses your path ("saber-tooth cat" would be more accurate, but "tiger" is more dramatic so let's stick with that). If your body didn't immediately release a flood of stress hormones, you'd stand there looking stupid rather than reacting quickly—grabbing your spear (fight) or climbing the nearest tree (flight)!

We must assume that the saber-tooth tiger also experiences a fight-or-flight response. It has never been established whether humans really were a meal of choice for saber-tooth tigers. After all, humans were "predators" too, and an encounter like this may have been a meeting of two hunters who respected one another. In any case, the fight-or-flight response is older than the human race, an alarm system installed in many creatures. And how does this alarm system work? Through molecules, of course.

The molecules lying dormant in our bodies first need to be roused by some sort of trigger. In the Stone Age, this might have been a saber-tooth tiger; today, it's Matthias's monster alarm clock. The clock's acoustic signal sends a nerve impulse from the brain to the adrenal glands via the spinal cord. Along with the pineal gland, the adrenal glands are among our bodies' most important hormone factories. This nerve impulse causes the adrenal glands to release what is probably the best-known stress hormone—adrenaline, which is promptly pumped into the bloodstream and makes its way to various organs. A hormone is nothing more than a messenger substance, a molecule that carries important information. And in this case, the message is PANIC!!

Adrenaline

While adrenaline is rushing through the bloodstream—and disappearing just as quickly—another hormone is gearing up for the stress war. ACTH (adrenocorticotropic hormone) is produced in the pituitary gland and travels through the bloodstream to the adrenal glands, the base camp for the fight-or-flight battle.

As soon as it arrives, ACTH unleashes a whole chain of chemical reactions. I like to picture it like an epic movie battle scene. Adrenaline is the forerunner who raises the alarm, while ACTH is the army commander who raises their fist and lets out the first battle cry, mobilizing the army and setting the carnage in motion. Finally, cortisol enters the bloodstream and makes its way to various organs as well.

Hormones can trigger a variety of physical reactions. Symptoms of a fight-or-flight response include an accelerated pulse, greater blood circulation in the muscles (RUN!!!), reduced blood circulation in the digestive system (drop everything, we have more important things to do!), deeper breathing, dilated pupils, sweating, goose bumps, and heightened awareness.

All these physical reactions to the release of my stress hormones mean that I am now, of course, wide awake, but the feeling of mortal danger isn't exactly pleasant. I can't blame this on molecules. Our bodily chemistry is designed for survival. The poor stress molecules don't know that Matthias's alarm isn't threatening my life. They just want to help.

The problem is that our modern world is full of stress—at school, at work, in our relationships. But very few situations are actually life-threatening, at least not acutely. Chronic stress definitely has an effect on our health. Luckily, to ensure that we and our molecules don't crack up completely, our stress system has a negative feedback loop that makes sure the body doesn't totally escalate and work itself into a panic. Among other factors, this is down to cortisol, the stress hormone with self-discipline. While adrenaline charges through the bloodstream once and then quickly disappears, cortisol stays in our system a little longer, ultimately inhibiting the release of ACTH and thus the production of cortisol itself.

FOR CONTRAST, LET'S look at a chemically perfect morning. While I snooze, the sun's first rays shine through my eyelids onto my retina, which is connected to the brain via the optic nerve. In the brain, the production of the sleep hormone melatonin is now inhibited in the pineal gland. The pineal gland is indirectly connected to the optic nerve and is sometimes referred to as the "third eye." This might sound esoteric, but there's something to it. In amphibians, the pineal gland is directly sensitive to light and really does act like a third eye.

While my melatonin level slowly decreases, a pleasant amount of cortisol is released. Ideally, I will wake up of my own accord.

When it comes to sleep, Matthias is unbelievably sensitive to light, so he always wears a mask. Because he blocks out the daylight completely, his melatonin level doesn't drop as quickly in the morning. Artificial darkness is just as confusing for our circadian rhythm as artificial light. Our modern world has plenty of both, which upsets our body clock. My hypothesis is that Matthias wouldn't need such a horrible alarm clock if he simply stopped wearing his sleep mask. Matthias thinks that his melatonin system is simply too sensitive and that he wouldn't get enough sleep without this quilted thing across his face.

What hampers both our arguments is that melatonin may not actually be a sleep hormone. For example, nocturnal animals also experience an increase in their melatonin levels at night—which would make it more of a "wake-up hormone." Laboratory mice often produce little melatonin at all due to a genetic mutation, and yet their sleep is perfectly normal. Plot twist! So does that mean melatonin doesn't make us tired? Well, on the other hand, many studies have shown that melatonin helps to treat insomnia and chronically delayed sleep. Hmm. So what now? Sleep researchers are yet to agree on the exact link between melatonin and sleep. As long as it remains unclear whether melatonin really makes us tired, Matthias and I can carry on debating the usefulness of his sleep mask.

Now, before you read the rest of this book, there's something I really need you to know: if you want to understand science, you need to lose the habit of looking for simple answers. This might sound arduous at first, but I promise that scientific thinking doesn't make the world drier; in fact, it makes it more colorful and literally full of wonder. So let's start by agreeing that melatonin isn't a "sleep hormone," but more of a "night hormone" that translates what the eyes can see (encroaching darkness) into the body.

A long-term experiment could shed some light on our personal melatonin dispute (and on Matthias's retina). Unfortunately, experiments with two participants are not statistically viable, so debate remains our only option.

I GO INTO the kitchen to make myself a coffee. When you get up, you should ideally wait an hour before drinking your first coffee—your morning boost of cortisol is your body's own way of waking up. Caffeine also encourages the body to produce cortisol. Great, you might think, I'll simply up my morning cortisol level with a coffee! Unfortunately (or luckily) our bodies don't work like that; they like balance. Bear in mind that, as time goes on, your body will acclimate to that coffee boost by reducing its own morning stress service. So it's better to wait until your body's own cortisol boost has leveled out again—which takes about an hour—before adding coffee into the mix.

Right now, I feel like all my morning cortisol has been wiped out in the space of a minute. I reach for the coffee to fight the tiredness I can already feel creeping up on me.

So, provided you're not feeling too hot already, grab yourself a coffee, tea, or hot beverage of your choice to drink as you read. There's nothing better than a hot drink to help you see the world in molecules. If I put my steaming cup of coffee on the table in front of me, before long the part of the table under the cup will warm up too. And if I wait even longer, the coffee will eventually go cold. Have you ever asked yourself where the heat actually goes?

This puts us right in the middle of one of my favorite topics, the *particle model*. It might not sound particularly exciting at first, but I guarantee you'll be fascinated. According to the particle model, *every substance in the universe is made up of particles*. They might be atoms, they might be molecules—conveniently, the particle model doesn't even need to know what these particles look like. Despite this extremely simplified way of looking at things, we can use the model to describe some parts of our world amazingly well—like my coffee, for instance.

So when I drink a coffee, I'm drinking coffee particles. Or tea particles, depending on your beverage of choice. Let's imagine that these particles are like tiny balls invisible to the naked eye. In reality, they are mainly water molecules with a bit of caffeine and a few other molecules like flavoring agents. These particles are constantly moving. You can actually see this movement, despite not being able to see molecules with the naked eye.

But how? Simple: take a glass of tap water and add a drop of coffee (ink works even better, but if you're drinking coffee anyway...). Even if the glass is on the table, not moving, it's only a matter of time before the drop of coffee disperses throughout the water, even if you don't stir it. Watching this happen might not blow your mind, but think about what's actually happening in this tranquil glass of water. It's a buzzing party, a chaotic glass of swarming, wriggling particles! And I'd like to invite you to this invisible particle party—this is where the chemistry begins.

HOME EXPERIMENT NO. 1

Particle party

Add a drop of coffee or ink
to a glass of water

Coffee or ink spreads
through the water

Incidentally, the glass, the coffee cup, the table, the floor on which it stands, the air—and, of course, you and I—are all made up of particles. And they are moving too! It's practically impossible for them to be still. At this precise moment, a particle party is taking place wherever you look—in your cup, beneath your feet, and in your body—it's just that you can't see it.

Now, you might ask what the point is of imagining a world made up of lots of tiny particles when we can't even see them (aside from the fact that it's just a cool idea; at least, I think it is). The point is this: you can use it to explain, for example, how the different *states of matter*—solid, liquid, and gas—are formed. Whether a substance is a *solid, liquid,* or *gas* depends on how much the particles are moving.

My coffee cup is solid because its particles only move a little; they are connected to one another via molecular bonds. We'll discuss chemical bonds in detail later on, but for now, let's compare the molecular situation to a concert with a tightly packed crowd. You can barely move at all, but still you jump around as best you can. This is a good representation of the particles in a solid object like a coffee cup.

In the liquid inside the cup—the coffee—the particles are a little more mobile, even if there is still a high level of interaction. At the aforementioned concert, we would now be in the mosh pit right in front of the stage, where everyone goes wild and throws themselves around. The gaseous air molecules, the ones we are breathing, are the wildest of all. They move without any regard for their fellow molecules. You would have to make the concert venue a couple of times larger so that everyone in the audience could run around freely and turn somersaults without getting in each other's way.

Water shows us that to switch between states of matter, we have to change the temperature. If we heat up solid water (ice), then it melts into a liquid; if we continue to heat it, the water vaporizes and becomes a gas. If the water vapor then hits a cool surface, like a bathroom mirror, it condenses and returns to liquid form. If we cool the water further, it freezes into ice.

Why am I telling you something you already know? Because I'm going to blow your mind. *Temperature is nothing more than the movement of particles. The hotter, the faster; the slower, the colder.* Isn't it cool to have a molecular definition of temperature? Isn't that much more satisfying than reading the temperature from a thermometer?

If you look back at your steaming cup of coffee, everything makes a lot more sense. The coffee is hot, which means that the water molecules are moving quickly and bumping into one another. The molecules that are evaporating are so quick and need so much space that their sheer need for movement makes them leave the cup and enter the gaseous phase.

So how does the heat of the coffee transfer to the cup and from there to the kitchen table? *Heat conduction* works through particle collisions and the transfer of kinetic energy. The coffee particles speed around the cup, constantly bumping into the sides. Like bumper cars, the cup particles also start to move more and vibrate faster. The cup particles collide with the particles in the kitchen table and cause them to vibrate faster as well. And because heat always conducts toward whatever is colder, the table grows warmer underneath the cup.

Now we can also see why the coffee eventually goes cold—for the same reason that a pendulum eventually stops moving. As with bumper cars, the particles slow each other down with every collision until they all return to room temperature—or "room speed."

All particles, and the universe as a whole (with everything it contains), follow the *first law of thermodynamics*. Here, we can compare this with the law of conservation of energy, which states that energy cannot be created or destroyed, but can only be transformed. To put it another way, the total amount of energy always remains the same. If a particle increases in energy, this same amount of energy must be lost somewhere else. If a particle transfers some of its kinetic energy to another particle when they collide, and the other particle becomes faster as a result, then the first particle must slow down. If this didn't happen, it would be as though energy had been created from nothing, and that isn't possible. The destruction of energy also goes against the laws of thermodynamics; some physicists and chemists get really annoyed when people talk about "wasting energy" (if you know any physicists or chemists, give it a try).

BEFORE WE CONTINUE through my day, I have one last thought experiment involving the particle model, perhaps the most interesting one of all. Touch the objects around you—some will feel warmer, others colder. But every object in a closed room has the same temperature, *room temperature*. So why does a metal spoon feel colder than a wooden table?

Well, one thing in the room isn't at room temperature, and that's your body. Your body temperature is higher than room temperature—at least I hope so, for your sake. What you feel when you touch a spoon or a wooden table is, in fact, your own body heat! If this heat is conducted away from you quickly, the object feels cold; if it's conducted away slowly, the object feels warm.

When I pick up the spoon, my hand particles collide with the spoon particles and cause them to vibrate. The faster the spoon's metal atoms vibrate, the warmer the spoon becomes. Metal is a good *heat conductor*; when the metal particles make contact with my finger particles, the movement travels well through the spoon. Metal is a good heat conductor because of its chemical bond (we'll take a closer look at this in Chapter 8). For now, picture the bonds in the metal as a jungle gym made from ropes. If a child starts climbing and jumps onto or shakes one of the ropes, the movement will quickly spread throughout the whole jungle gym. If there's another child on the other side of the jungle gym, they'll be shaken as well. At the same time, the movement of the jumping child is curbed by the law of conservation of energy. When the child transfers their kinetic energy to the ropes and to the other child, they themselves become slower; their movement is curbed. In terms of thermodynamics, this means that the child becomes slower, with less energy—so colder.

The jungle gym could also be made from solid poles. If the child jumps onto a pole, it won't have much impact on another child on the same structure. The child's own movements will barely be curbed or transferred elsewhere, so the child will become faster and warmer. This kind of jungle gym is like a poor heat conductor, such as wood. If you place your hand on a wooden table, only the wood particles in direct proximity to your hand will vibrate. The vibration and movement don't travel so well through the rest of the wood, so it feels warmer than the metal spoon.

IF TEMPERATURE IS nothing more than the movement of particles, this makes it easier to visualize the *second law of thermodynamics*, which states that heat always flows from something warm to something cold and never the other way around.

If you put a bottle of cola in an ice bucket, the cold from the ice doesn't flow into the bottle. In fact, the exact opposite happens—the warmth from the bottle flows into the ice cubes, which warm up, and this is how the bottle cools down.

Now that you know this, listen out for the next person who says "Close the window, the cold's getting in." Instead of leaving this thermodynamic baloney unchecked, say "I think you mean the heat's getting out!" And if it starts to bug you when people talk about "wasting energy," then you'll fit right in with the nerds. Congratulations, your introduction to *physical chemistry* is complete—hopefully before you've even finished your coffee!

MATTHIAS COMES INTO the kitchen and strokes my hair apologetically.

"Sorry, I forgot to tell you I was going running today."

"It's fine," I say. "I need to adjust my sleep pattern again anyway."

Although I know better in theory, I love to sleep late on weekends, which leaves me with "social jet lag." Obviously, my circadian rhythm can't tell the difference between weekdays and weekends. Weekends are great, but they are a modern social construct that means nothing to our bodies. Our natural melatonin level is more or less based on the sun. And yet I'm exhausted when the sun rises and end up going to bed far too late. My life of coffee, artificial light, and awful alarm clocks is constantly confusing my body with false stimuli. Researchers have observed that, after a week of camping with no coffee, artificial light, or cell phones, a person's melatonin cycle readjusts to solar time. It's just a shame I don't like camping.

One thing is truly strange—theoretically, our biological clock works without light too. On this planet with twenty-four-hour days, our biological clocks have evolved to follow twenty-four-hour days as well, with only minor variations. The light helps us to set our biological clock—to synchronize the days—and to adjust to jet lag, for example.

In 2017, the Nobel Prize in Physiology or Medicine was awarded to three American researchers who revealed the workings of our biological clocks. They kept fruit flies in two different chambers labeled "New York" and "San Francisco" with staged lighting based on the solar rhythms of these two coastal cities. The fruit flies were repeatedly "flown" in an "airplane" (a glass) to the other "city." The researchers observed how the flies coped with the three-hour time difference.

They discovered that two different genes are essential to our biological clocks. Genes are where chemistry gets really exciting! Our DNA not only is a molecule in its own right, but also ensures the production of other molecules vital to life. All the information we need to live is coded in our genes, including information about our biological clock. This code can be read and translated when these genes produce *proteins*. In other words, the genes have the plan, and the proteins implement that plan (proteins are highly shrewd molecules that will come up throughout the book).

So the two "clock genes" produce two "clock proteins." Initially, both proteins increase in concentration, but then they combine to form one unit. Teaming up allows them to carry out the plan made by the genes—to inhibit their own production. That's right, these proteins are produced to stop their own production, to stop their own genes from being "read." Similar to cortisol and stress, what we have here is a negative feedback loop. If no more clock proteins are produced, their concentration decreases further. Eventually, the protein concentration drops so low that the reading of the genes can no longer be restricted—and protein production starts from scratch. This whole cycle takes almost exactly twenty-four hours. Day and night are coded in our genes.

I get the feeling that something isn't right with my genes. I'm convinced that my body is designed for a thirty-hour day—I need much longer days and a lot more sleep. I'd like to get myself checked out.

"I HAVE TO go," says Matthias.

My cell phone vibrates. I'm surprised to see a message from Christine. She's awake already?

"I think Jonas is dead," she writes.

"I'll call in a second," I reply.

Already dressed for running, Matthias sticks his head around the open door and asks if he needs to take a key.

"No," I say. "Close the door, the heat's getting out!"

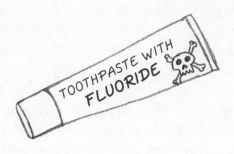

Death by Toothpaste

"**WHERE ARE YOU?**" I ask Christine when she finally answers her cell.

"On my way to the lab." She sounds annoyed.

"So what's going on with Jonas?"

"I've just left his place," she huffs.

"So you did spend the night with him? How—"

She interrupts me. "Mai, he uses NATURAL TOOTHPASTE."

"What?"

"Without fluoride."

Oh crap. Jonas is a really sweet physicist Christine has been seeing for a few weeks. We've actually known him for quite a long time through our friend Hannes, who's also a physicist. Although Jonas is extremely good-looking,

Christine was never particularly interested in him. I would describe her as sapiosexual—she's only attracted (emotionally and physically) to intelligent people. So when Hannes told us that Jonas was "a real brainiac" and came top of the class every semester, Christine was suddenly hot for him. Which makes it even more alarming that he uses toothpaste without fluoride.

"Are you sure?" I ask. "Maybe it was just the tube; they all try to market themselves as organic these days. I mean, there are herbal toothpastes that contain fluoride."

"No, it clearly said 'NO FLUORIDE.' And I read the list of ingredients as well."

"OK. So what did it contain if there was no fluoride? What was the substitute? Did you—"

"That's not the point," Christine interjects.

Oh dear. It must be serious if she doesn't even feel like ruminating over the ingredients.

"I'm so turned off. I think Jonas is dead to me."

Death by toothpaste, I think. Ironically, this is exactly what Jonas is afraid of.

"But did you ask him about it? Maybe he wasn't paying attention when he bought it."

"He says that fluoride calcifies the pineal gland. But he didn't even know where the pineal gland was!"

Well, I think to myself, physicists are no chemists.

"Hydroxyapatite," Christine says suddenly.

"What?"

"The fluoride substitute in this herbal toothpaste," she snorts. "Ridiculous."

"You mean hydroxyapatite as in tooth enamel?"

"Yes! How is that even allowed?"

"Interesting," I say.

"Please make a video about it," says Christine. "I'm at the lab now. Speak to you later."

Thinking about it, a video about fluorides and toothpaste would be a great idea. Many people find it odd that I studied chemistry and completed a doctorate only to end up doing "something in the media." But I do it with conviction. Laboratory research isn't the only way for a scientist to help humanity. Talking about science is just as important. It's pretty damn difficult for laypeople to find scientific information that is both understandable and correct. The internet is full of half-truths and lies sold in an alarmingly persuasive manner. Specialist books might contain reliable information and scientific journals might present the latest research findings, but they're a nightmare even for experts to read—especially the journals. Science is like an elite society with a secret code. While it makes sense for experts to use specialist language among themselves, it's absurd that nonexperts can't understand what they're saying; a large proportion of research is financed through public funds, but taxpayers can't figure out how their money is being spent. I actually think we need more scientists on YouTube and television to "translate" this language.

WHILE I FINISH my breakfast, let's start with the difference between fluoride and fluorine—the perfect topic given that I'm cooking my eggs in a Teflon pan. Keep that in the back of your mind while I explain.

Fluorides are a form of the element *fluorine*. If you look at the periodic table (see the back of the book), you'll find fluorine (F) in the seventh main group, known as the *halogens*. Fluorine is a gas with an odor similar to *chlorine*, the halogen used in swimming pools. I hope you'll never smell it though—fluorine is pretty damn dangerous.

What do I mean by "pretty damn dangerous"? I mean that even the smallest quantity of fluorine gas in the air would burn your eyes and lungs. Fluorine acts aggressively because it's highly reactive. The general rule is that the easier and faster a substance enters into a chemical reaction with another substance, the more dangerous it is—because it's less easy to control. There are other reasons why substances can be dangerous or poisonous, but we'll come to that later.

In any case, fluorine gas reacts with water to produce *fluoric acid*. "Acid" sounds scary, and rightly so; if you accidentally poured some on your hand, it wouldn't just burn your skin, it would eat right through to the bone and dissolve it. Other dangerous acids like *hydrochloric acid* (the equivalent for chlorine) seem practically harmless in comparison.

So please keep away from elemental (pure) fluorine and fluoric acid! Don't worry, there's nothing you need to do. Luckily, neither substance occurs in nature (or in toothpaste) thanks to another simple, general rule of chemistry: the more reactive a compound, the less common it is in nature. It's logical really—if fluorine is so aggressive that it reacts with everything it meets, we can assume that all the fluorine out there has already "reacted itself away."

However, you can produce fluoric acid in a laboratory—not because you're a mad chemist who wants to take over the world, but because of your fried eggs. In a chemical laboratory with the right equipment, you can pair fluoric acid with whatever you want. If you choose the right substance, you can produce polytetrafluoroethylene, PTFE for short, also known as *Teflon!* And now we're back to my pan and eggs.

What about the fluorine atoms sitting in my Teflon pan? Is there fluorine on my eggs? Good question—let's dive in.

Most elements, even the reactive and aggressive ones, have a stable form in which they are unreactive and relaxed. How an atom is made up on the inside determines whether it tends to be aggressive or relaxed. As in life in general, in chemistry it's what's inside that counts. (Well, almost always. In the particle model, what's inside the particles plays no role.)

We often imagine atoms to be the smallest particles, the smallest components of our world—but that's not true. Atoms are made up of three different elementary particles: *protons, neutrons,* and *electrons.* Protons have a positive charge,

neutrons are electrically neutral, and electrons have a negative charge. Our entire world, in all its diversity, is made up
of just three different building blocks (a physicist would say
something different at this point, but let's not make things
unnecessarily complicated). It's astonishing really. When I
mix eggs, flour, and milk and heat them up, I might end up
with a pancake or spaetzle (a type of Southern German pasta),
depending on how I combine them. And while these may be
two different foods, they have much more in common than
gold and oxygen, for example. And yet gold (a metal) and
oxygen (a gas) are constructed from the same three building
blocks. Isn't that incredible?

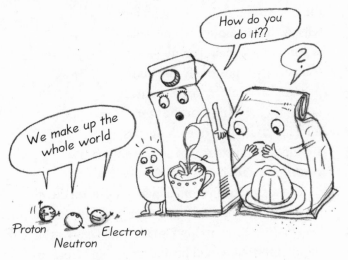

What makes gold gold and oxygen oxygen if not the building blocks themselves?

The type of element is determined by the number of protons. The
periodic table tells us how many protons an element has.

And how are the elements arranged in the periodic table? By their *atomic number*, which equates to the number of protons. A quick look at the periodic table tells us that oxygen is number eight, so it has eight protons. Gold is number seventy-nine, so it has seventy-nine protons. And it is this difference alone that makes oxygen oxygen and gold gold.

Long ago, alchemists (who came before chemists) tried to transform nonprecious metals into gold. Today we know that, due to the structure of atoms, no laboratory technology can make this work.

You've probably seen images like the one to the right. This diagram tells us that an atom has a nucleus (a core) and a shell. The nucleus is made up of positively charged protons and electrically neutral neutrons, meaning that the atomic nucleus has a positive charge. The atomic shell is made up of negatively charged electrons that circle around the nucleus.

The weight of an atom depends solely on the nucleus (the number of neutrons and protons). Electrons weigh practically nothing, so it's not even worth including them—that would be like weighing elephants with feathers on their backs and insisting on counting the feathers.

Naturally, a single atom weighs very little because it's so small, but atoms do of course have a mass—otherwise this book (or your body) wouldn't have any mass either.

Mass of a carbon atom: 0.00000000000000000000002g

With seventy-nine protons, a gold atom is much heavier than an oxygen atom with eight protons. Then we have the neutrons, each of which weighs about as much as a proton. As a general rule, each atomic nucleus has roughly the same number of neutrons and protons. Add them all together and a gold atom is around twelve times heavier than an oxygen atom.

In contrast, the volume (size) of an atom is determined not by its nucleus, but by its electron shell. Compared to the atom—which is tiny in any case—the nucleus is so small that its volume can be ignored. It's like cotton candy on a stick. If the cotton candy is the electron cloud, then the stick is the atomic nucleus. The size of the cotton candy depends solely on the cotton candy itself; ultimately, the thickness of the stick makes no real difference. Imagine the atomic nucleus like a "point mass," a mass that has practically no volume and is concentrated on one tiny point.

So the atom is as large as its electron shell. The exact size of this shell depends (among other things) on the number of electrons. Conveniently, an atom has the same number of electrons and protons by default. The positive and negative charges cancel each other out, and so we have an electrically neutral atom. This means there are seventy-nine electrons buzzing around in the electron shell of gold, and just eight in oxygen. And since every electron needs its space, gold has a larger electron shell than oxygen. Therefore, a gold atom is more than double the size of an oxygen atom.

NOW THAT WE'VE dealt with the mass and volume of atoms, let's turn our attention to the truly fascinating stuff—their chemical properties! At this point, we're going to forget about the atomic nucleus, which has nothing to do with chemical reactions. Chemical reactions only take place in and between electron shells, which is why you can't turn iron into gold— you would have to add protons to the iron nucleus, and that isn't possible. It's not so easy to change the number of protons in a nucleus (with the exception of radioactivity, in which heavy, unstable atomic nuclei gradually decay). And this is why it's worth paying a bit more attention to the electron shell—now things get really exciting!

In the atomic model on page 26, we saw how electrons circle around the nucleus. But that was a very simple model. As with the particle model, remember that *a model never describes reality; it merely presents a simplified view.* Models only ever apply in certain circumstances (just like fashion models who are chosen for their stunning looks or are skillfully retouched— they don't represent the average human being). As you'll have noticed, I love simple models. Why make things more complicated than they need to be?

The following model—the *shell model*—is pretty useful for understanding chemical reactions:

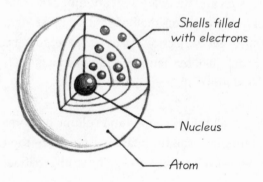

Shells filled with electrons

Nucleus

Atom

According to the shell model, electrons can't circle around the nucleus however they like, but only at very specific distances. Think of these distances as layers of onion skin surrounding the nucleus (I would have called it the "onion model," but nobody asked me).

Like the way the Earth and the other planets orbit the sun at specific distances, electrons may only circle around the nucleus at a certain point. But why are some distances forbidden? It's something to do with quantum mechanics; particles as tiny as electrons are subject to the rules of quantum physics, not traditional physics.

It's hard for us to get our heads around quantum physics, because everything we see and experience follows the rules of traditional physics. Delving into quantum mechanics is a bit like imagining a color we've never seen. So let's use the following analogy instead:

The shells are like fixed rows of seats in a cinema. You can only sit on the seats, not between rows. (And why would you? You wouldn't see anything.)

So who sits in which row? In each element, the electron shells are filled from the inside out. Once one shell is full, the next one gets its turn. The electrons in the outermost shell, the *valence electrons*, are particularly important. Because they are located so far from the nucleus, they have special properties; the farther the shell is from the atomic nucleus, the weaker the attractive force between the positively charged nucleus and the negatively charged electrons. The nucleus has a much weaker hold on the valence electrons than it does on the electrons in the inner shells. While the inner electrons are inert and like to stay close to their positive nucleus, the valence electrons are extroverted, highly reactive souls who like to take part in chemical reactions.

Apart from the fact that the valence electrons are farthest from the nucleus, there is something else that makes them restless; while the inner electron shells are full, many elements have a partially empty outermost shell. The number of electrons is limited; as I mentioned previously, it matches the number of protons. Now, I don't mind having a whole row to myself when I go to the movies, but then I'm not an electron. Electrons—particularly the extroverted valence electrons— hate empty seats! This is the odd thing about atoms; they really want their shells to be full.

This desire is expressed in a curious way. Elements with a few empty seats in their valence shell are not so aggressive. But if there's only one empty space to fill or the valence shell contains just one solitary electron, things start to get uncomfortable. Elements that have narrowly missed out on a full shell are the most aggressive. It's like the World Cup, where the runners-up often cry the most—close but no cigar.

So, the chemical temperament of an atom is linked to how many valence electrons it has. Our fluorine has seven, but its valence shell has space for eight. And that one empty spot makes the fluorine furious. It restlessly searches other atoms and molecules for the missing eighth electron and won't stop until it fills its final spot.

Fluorine isn't the only element with this problem. *Almost every element in the main groups of the periodic table would like to have eight valence electrons.* This is known as the *octet rule;* the name is a bit misleading because it's not a fixed rule (like a law of physics), but simply a model—although admittedly a very practical model closely connected to the shell model. The octet rule can be used to clarify not only which elements are particularly reactive, but also which reactants go well together. Every element has needs, and fulfilling the octet rule may be one of them. Chemical reactions and bonds satisfy these needs (much like in humans, actually).

Fluorine is like a baby crying for food. Once fed, it's quiet and peaceful (parents may argue otherwise...). Once fluorine forms a new bond that provides its longed-for eighth electron, nothing much happens.

So what does this mean for my pan? Teflon is produced by combining fluorine with carbon; one of carbon's properties is to generously share electrons and shells with other atoms (the exact form this bond takes will be explained in Chapter 8). It would take a whole lot of energy to extract the fluorine from this happy state of affairs. To break the bonds, you would need to heat Teflon to over 360°C (680°F); the recommended maximum temperature for Teflon pans is 260°C (500°F), and the optimum temperature for my fried egg is around 83°C (181°F)—any higher and the egg white solidifies.

So, according to the octet rule, the fluorine and carbon atoms in my pan have achieved everything an element can possibly achieve: a full valence shell. The chemical bond between fluorine and carbon is truly a model marriage—neither party has eyes for any other atoms or molecules. Not even for the attractive proteins in the fried egg currently sizzling in my Teflon pan.

If food sticks to the pan, then the pan molecules and food molecules are simply interacting. Teflon has no interest in my fried egg or any other food. The fluorine in my pan is probably relieved to have left behind its restless youth as fluoric acid. "I have everything I want, leave me in peace," it thinks. "Don't mind me, I was never here," my fried egg replies, sliding onto my plate without leaving a trace.

As you read this whole discussion of Teflon, were you thinking to yourself, "Whoa! What about all the headlines saying Teflon cookware is dangerous to use? Why is Mai cooking with a Teflon pan?" Well, I use Teflon cookware because when it's used *correctly*, it's safe. Remember those temperatures I mentioned above? It's *overheating* a Teflon pan that's a bad idea. The gases released by a severely over-heated Teflon pan can cause flu-like symptoms. I talk more later about misleading headlines in science journalism (in Chapters 3 and 4).

By the way, the fluorides in toothpaste are just as contented as the fluorine atoms in the pan. I previously mentioned that an uncharged atom contains the same number of electrons and protons because the electrical charges cancel each other out. But an atom can have a charge, and then it's known as an *ion*. A *negatively charged ion is called an anion* and develops when there are more electrons than protons. In chemistry, the names of anions end in *-ide*. Inversely, *a positively charged ion is called a cation* and develops when there are fewer electrons than protons. There is no special word ending for cations. However, the -ide on the end of "fluoride" tells us that this is a negatively charged ion; someone has given fluorine its longed-for electron. A singly charged anion with eight valence electrons, fluoride has now fulfilled the octet rule and is thoroughly satisfied.

How does this happy situation come about? One hot contender and reactive partner is the element *sodium*, a member of the first main group of the periodic table and the *alkali metals*. We know sodium (Na) from sodium chloride, or cooking salt, but this doesn't contain any elementary sodium. You've probably never seen pure sodium; there is no elementary fluorine or elementary sodium in nature.

Sodium is a glittery gray metal so soft it can be cut with a knife. Might sound nice, but sodium reacts intensely if you put it in water (there are videos on YouTube, but don't try it at home!). This means sodium is aggressive and the perfect partner for fluorine. The sodium atom isn't missing an electron; instead, like all alkali metals, its valence shell contains a solitary electron that would much prefer to leave the atom than continue hanging around by itself. Sodium wants to get rid of this electron, and quickly—so if something like fluorine comes along, great! They can both fulfill their octet rule. Together, they produce the salt *sodium fluoride*, the stuff that's in our toothpaste (the same principle applies to sodium and chlorine and our cooking salt, *sodium chloride*).

So the fluoride in toothpaste isn't particularly reactive, but unreactive doesn't automatically mean it isn't poisonous. Could it be poisonous? Death by toothpaste? And why is the fluoride even in there? I need to brush my teeth now, so let's continue this discussion in the bathroom.

3

Down with Chemism!

EVERY BATHROOM IS a chemistry lab, or at least a chemical cabinet. My nonchemist friends don't like that comparison— it makes them think of poison—so I have to be cautious when trying to pass on my love of chemistry. Although elemental (pure) fluorine and sodium are highly aggressive when combined, the word "chemical" isn't inherently negative. Whether poisonous, healthy, or crucial to our survival, every substance in this world is in some way chemical!

Chemistry may not necessarily be poisonous, but it's pretty infectious. My dad is a chemist and so is my brother. My best friend Christine is a chemist, and I married a chemist as well. And we're all totally normal, I swear!

My dad worked in hair cosmetics research for quite some time. We would stroll through drugstores and read the ingredients listed on the bottles. Sometimes we'd see one of the substances he had developed. My dad is also the reason why I became a polymer chemist. Some cynical chemists would say that polymers are plastic, but that's a shamelessly reductive definition. Teflon, polytetrafluoroethylene, is a polymer too, and it's also possible to produce biologically compatible polymers—for example, to transport cancer medication in the body or as the basis for artificial organs. Just plastic? Please. You must be joking!

Polymers are long-chain molecules. They are made up of many small molecular units, or *monomers*, that bind together to form long chains. *Polysaccharides*, or carbohydrates, are also polymers. So they don't have to be artificial at all; they can be found throughout nature. Wood and plant fibers are made up of cellulose fibers—and, yes, these are polymers too. Just like our DNA. And what's cool is that we can make polymers in the lab as well.

To give one example, my dad used to develop polymers for volumizing hair sprays and conditioners that targeted split ends. This alone made chemistry seem interesting to me. As a female chemist, I'm surprised that chemistry is still stereotyped as a masculine domain. Sometimes people ask me why I'm on YouTube—isn't it just full of makeup tutorials? I don't see how people can be interested in cosmetics but not in chemistry. Even the production of "natural" cosmetics—soaps and products with naturally occurring ingredients—requires an understanding of chemistry.

MY TUBE OF toothpaste is almost empty. I squeeze the last of it onto my toothbrush while my fried egg, bread, coffee, and orange juice sit happily in my stomach, waiting for my metabolism (a veritable frenzy of chemical reactions) to work its magic. And there's plenty of chemistry going on in my mouth too. The *sugar* in the bread and orange juice triggers a particularly interesting reaction. Incidentally, orange juice contains just as much sugar as cola. And even bread is ultimately made of sugars, or carbohydrates, which are nothing more than sugar polymers.

We're constantly eating sugar in various forms. Not (just) because we're greedy, but because our body converts sugar into energy. Our brain in particular runs on sugar, which is why it has us conditioned to love chocolate and gummy bears—not great when we see sweet treats everywhere we go.

We're not the only ones who love sugar; the bacteria and microorganisms that live on our teeth love it too. As you read this, hundreds of different types of bacteria are roaming around your mouth—and mine too, obviously. Every time you kiss someone, you exchange millions of bacteria through your saliva. Sorry if that grosses you out, but as a chemist I enjoy looking at the world on a small scale and thinking about things that can't be seen with the naked eye.

The bacteria on our teeth live in plaque—the thin, watery layer that covers our teeth. Plaque is also known as tartar. Toothpaste and mouthwash brands like to say that they help "fight plaque." I don't want to be a spoilsport, but it's impossible to get rid of it completely. What you can do, however, is change the conditions inside the plaque to make life difficult for the bacteria that live there.

When we eat sugar or carbohydrates, the bacteria chow down and fart out acid in return. This might not be the most accurate analogy, but when I told this to a friend's five-year-old daughter, she found it hilarious and has apparently been much more enthusiastic about cleaning her teeth ever since (so I highly recommend it). Ultimately, the bacteria metabolize the sugar as part of a complex chemical process. Bacteria have a metabolism, just like we do, that they use (for example) to convert sugar molecules into acid molecules—and they do it right on the surface of our teeth.

Tooth enamel is largely made up of a mineral called *hydroxyapatite*. Aha! That's the stuff that replaces fluoride in Jonas's toothpaste. Cleaning your teeth with tooth powder is a truly strange idea and not particularly effective against tooth decay. This becomes clear once you understand what tooth decay actually is. The hydroxyapatite in our tooth enamel doesn't like acids, which cause it to dissolve. It dissolves very slowly but that's bad enough. However, sugar converted to acid isn't the only problem. Many foods (like orange juice) contain acids themselves. Sugar plus acids equals twice as bad for the teeth. Coffee is acidic too, which is why there's fluoride in my toothpaste. I want to stop my tooth enamel from dissolving!

In the previous chapter, we learned that fluorides are negatively charged ions, known as anions. Hydroxyapatite, the mineral in our teeth, also contains anions (hydroxide ions). But fluoride is small and gets almost everywhere, even in our enamel. It gets in when we brush our teeth and forces out the hydroxide ions. Sounds aggressive, but it's a good thing. This exchange creates an extremely thin layer on the surface of the teeth of a more solid, more stable mineral called *fluorapatite*, which can't really be touched by acids. Shark teeth are almost 100 percent fluorapatite, which makes them particularly solid and explains why shark bites are so painful.

Toothpaste with fluoride...

...throws out the hydroxide ions...

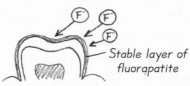

...and swaps in fluoride ions

Stable layer of fluorapatite

So how does Jonas's toothpaste work without fluoride? Basically, not so well. The fluoride has been replaced with hydroxyapatite, the mineral in tooth enamel. The reasoning behind this is to replace the enamel that dissolves with new enamel. But this doesn't form a protective layer against acidic plaque. Happy days for tooth decay and despair for Christine, while Jonas believes that fluoride calcifies the pineal gland.

Moving away from tooth decay for a minute, is fluoride poisonous?

As Paracelsus said—and this is very important—*the dose makes the poison.* (To be more precise, he is quoted as saying that "poison is in everything, and no thing is without poison. The dosage makes it either a poison or a remedy.")

You could definitely poison yourself with a large quantity of fluoride; the acute, lethal dosage for adults would be a few grams. I can't think of a single realistic scenario (outside of a chemical laboratory) in which a person would be exposed to such a quantity. Even if you took part in the world toothpaste-eating championship, you would probably vomit before reaching a lethal dose of fluoride. However, it is possible to overdose on fluoride over a longer period of time, leading to skeletal fluorosis and brittle bones. For example, employees in steel and ceramics factories work with substances that contain fluoride and, in the worst cases, could inhale them for years on end. In some severely polluted areas of the world (studies have been conducted in regions of China and in Mexico City, for instance), the water may contain higher quantities of fluoride, which is then ingested by residents over a long time frame.

So what about toothpaste with fluoride? The fluoride concentration in toothpaste is carefully examined to ensure it remains effective but noncritical. My tube of toothpaste says "1450 ppm F-," which means that every seven hundredth particle is a fluoride ion. We don't need any more than that; a few fluoride ions on the outermost surface of the teeth are plenty to protect against decay. Of course, the same concentration in drinking water would be far too high, but concentration is always a matter of context. Toothpaste is a topical treatment; we use a negligible amount and spit most of it out again. That is, unless you're a child who likes to eat toothpaste. Some kids will eat anything they can find; my brother liked to eat sand when my parents weren't watching. This is why children's toothpaste contains less fluoride—we shouldn't eat toothpaste, but kids can't always be stopped. In addition, teething children are particularly susceptible to dental fluorosis, which takes the form of spots on the teeth; in the best cases, these are white spots, but darker, unsightly spots are possible ranging in color from yellow to brown.

TO SUMMARIZE, FLUORIDE works well against tooth decay in the concentration permitted in toothpaste, while larger quantities carry the risk of fluorosis—and Jonas's fear of a calcified pineal gland opens up a whole different can of worms. This appears to be a surprisingly common worry, one of those strange internet phenomena where fears with no scientific basis are incubated and fueled by forums and Face-book groups. A Google search turns up all sorts of outlandish claims. "Fluorides calcify the pineal gland!" "Fluorides calcify the brain!" "Fluorides make you stupid!" While such articles may often link to scientific studies, a closer look shows that these fears are in no way confirmed. Scientific studies may be a sound basis for an argument, but problems can arise when nonexperts start bandying about scientific results. Scientific publications aren't written for nonspecialists or laypeople. They are a way for experts to communicate and exchange detailed information at the highest level of specialism. If these studies are not presented properly by science journalists, there's a serious risk of misunderstanding or even misuse by nonexperts.

Intensive, appropriate research shows that the fear of toothpaste calcifying the pineal gland is based on a few, extremely tenuous studies. The main culprit is probably a long-term study of pregnant women in Mexico City, where a high concentration of fluoride in the drinking water and the environment was combined with several other forms of pollution, including a high concentration of lead. Their children were later found to have a slightly reduced IQ, but only with major statistical variance. More experiments would be required before any well-founded suppositions could be made. However, in light of the generally high level of pollution, the results cannot be specifically attributed to fluoride (let alone to fluoride in toothpaste). Nevertheless, this study made the rounds with headlines like "Fluorides Reduce Children's Intelligence in the Womb." This is bad science journalism.

MY CELL PHONE vibrates again.

"You know the worst bit? Jonas hasn't been to the dentist for three years and can't remember the last time he had a cavity!!" Christine writes, ending her missive with an angry emoji. I can't help but laugh; I can just imagine how worked up she must be. If Jonas never gets cavities, then he can keep his toothpaste. Christine is just annoyed by his blissful ignorance. Some people are less susceptible to cavities than others. Personally, I definitely wouldn't go without fluoride; I'd develop cavities in an instant. It may simply be that some people have a different type of plaque that can be tackled with a dental routine focused on neutralizing the pH value. In addition to fluorides, toothpaste has other important components such as surfactants, which are basically soap (more on that in a moment), and tiny particles that act as abrasive agents, like in scouring cleaners. After all, you want to get rid of all the leftover food too. If you can manage without fluoride *and* don't get cavities, then do what you want. But if you suffer from cavities *and* a fear of fluorides, do yourself a favor and put a stop to decay by using normal toothpaste that contains fluoride.

I RINSE MY mouth and get into the shower. I find myself wondering how awful humans would smell if we didn't shower so often. We'd smell as bad as skunks. But if the smell didn't bother us, there wouldn't be much point in showering, at least not as frequently as most people do today. You might be surprised to learn that daily showering not only is unnecessary, but can actually be harmful. Why? To answer that, we need to understand both our skin and our shower gel a little better.

Like our tooth plaque, our skin is densely populated by a variety of microorganisms. As unpleasant as it might be to imagine bacteria and other organisms crawling around on us at any given time, this microbiome is generally harmless— and even useful. Think of the skin and its inhabitants as a complex, well-balanced ecosystem.

However, we also come into contact with unpleasant microorganisms like germs, particularly via our hands.

Although our skin doesn't let any of them through, the germs get into our bodies when we rub our eyes or handle our food, for example. This is why it's important to wash our hands with soap, and this brings us to perhaps the most important part of bathroom chemistry: *surfactants*.

I mentioned before that surfactants can also be found in toothpaste, but the classic surfactant is soap, like in hand soap or shampoo. Washing with water would be far less effective without soap—our skin is *hydrophobic*, which literally means "water-hating." The membranes of our skin cells and the spaces between them are built from hydrophobic molecules. Hydrophobic substances can't be mixed with or dissolved in water. Other hydrophobic substances include *oils* and *fats*, so "hydrophobic" can also be replaced with the term *lipophilic*, which means "fat-loving." If you make a salad dressing from vinegar and oil, you will see that the water and oil don't mix; a "phase boundary" forms between them. The water molecules and oil molecules want nothing to do with one another; they repel each other and prefer to keep to themselves.

The opposite of hydrophobic is *hydrophilic*—"water-loving." Alcohol is a hydrophilic liquid, so it mixes well with water (luckily, otherwise we wouldn't be able to drink it). The ethanol molecules and water molecules interact well and attract each other. Sugar and cooking salt are also hydrophilic, which explains why they dissolve so well in water, but not in oil.

In principle, every substance can be categorized as hydrophilic or hydrophobic, but the distinction is fluid. Our skin tends to be hydrophobic—this is the best way for it to protect us, and we don't want it to dissolve in the rain or the shower.

However, this also means that it doesn't interact particularly well with water. In addition, our pores produce not just sweat, but also sebum (fat)—a hydrophobic substance. Bacteria also have skin; single-celled organisms have a cell membrane that is also hydrophobic. And because the sebum and bacteria really want nothing to do with water, they remain comparatively unimpressed when doused in the stuff. If you've ever tried to get a grease spot out of your clothes using water, then you'll know what I mean.

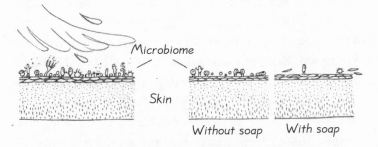

Microbiome

Skin

Without soap   With soap

Thousands of years ago, humans discovered *soaps*, or *surfactants*. These magic substances are *amphiphilic*, which means they combine hydrophobic and hydrophilic properties in one molecule. Traditionally, soaps are long molecules made up of two parts—a long hydrophobic tail with a hydrophilic head. Surfactants are popularly visualized as tiny pins; the needle is the hydrophobic part and the head is hydrophilic.

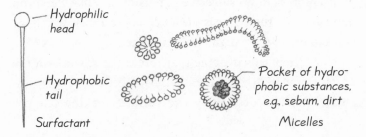

— Hydrophilic head

— Hydrophobic tail

Surfactant

— Pocket of hydrophobic substances, e.g. sebum, dirt

Micelles

If you put surfactants in water, fascinating things happen. The molecules form geometric structures all by themselves. This happens because the hydrophobic tails want nothing to do with the water and arrange themselves so that they have as little contact with the water as possible. This leads to the formation of *micelles,* in which all hydrophobic tails turn inward and all hydrophilic heads turn out toward the water. Micelles can be spherical or shaped like rods or worms.

Surfactants tend to arrange themselves on boundaries, a behavior known as *surface-active.* If you put olive oil and soapy water in a glass, most of the surfactant will position itself on the boundary between the water and the oil so that the hydrophilic heads are facing the water and the hydrophobic tails are facing the oil. The same thing happens on the boundary between water and air. Air isn't actually hydrophobic, but all the hydrophobic tails really care about is that it isn't water, so they orient themselves toward the air, just like a row of ducks (heads in the water, tails in the air).

This is also what makes bubble baths and soap bubbles possible. A soap bubble may be very fragile, but it's actually surprisingly stable—a hollow sphere in which the curved layer of water is subject to immense tension.

This sounds a bit weird at first. How can a liquid be subject to mechanical tension? When you try to make bubbles, for example, water is subjected to tension, just like when you bend a plastic ruler. This is the *water surface tension,* which we will explore in more depth in Chapter 10.

Until then, let's imagine that forces are at work on the surface of water that make it almost a little rigid. But now, surfactants are occupying the surface, making it more

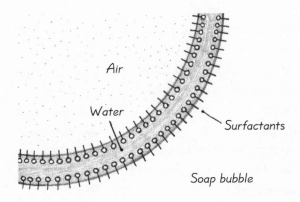

flexible and elastic, if you will. A rigid plastic ruler becomes softer, more elastic, easier to bend without risk of breakage. *Therefore, surfactants reduce water surface tension.* And this means that the water can easily morph into soap bubbles and even smaller, more curved bubbles without which we would have no bubble baths.

Because surfactants are amphiphilic, they are great intermediaries between hydrophilic substances like water and hydrophobic substances like sebum, dirt, and bacteria. When we wash ourselves with soapy water, hydrophobic substances on the skin can be trapped inside the micelles and then washed away with the water. The same principle applies to washing powder, household cleaners—and toothpaste. So mundane and yet so ingenious.

So how do we make these wonderful little pins?

The first soaps were made by cooking oils or fats with plant ashes. The underlying chemical reaction is called *saponification.* The base material is always fat. From a chemical perspective, fats and oils are *triglycerides*; a fat molecule is a fusion of three *fatty acids.* Fatty acids were always meant to form soaps, made up of a long hydrophobic tail with an *acid group* on the end that can serve as a wonderfully hydrophilic head. The pin shape is practically predestined. However, the acid groups within the triglyceride are bound together in such a way that they cannot interact with water. Imagine a fat or triglyceride as three pins bound together by their heads. The acid heads can be freed from their bond if an *alkaline* reactant is brought into play. Plant ashes contain (alkaline) potassium salts. And *acids and alkalis react with one another very easily.*

If we now heat up fat with potassium salts, the bond between the triglycerides will break, and we end up with free fatty acids with *saponified* acid heads. Saponified means that the acid group is now negatively charged—like a fluoride ion. And most charged groups get on famously with water (we'll find out why in Chapter 10). And this is how you get from a fat to a surfactant.

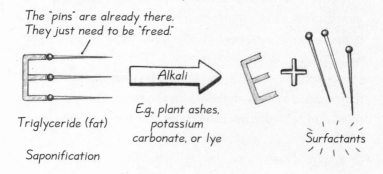

The "pins" are already there. They just need to be "freed."

Triglyceride (fat)

Alkali

E.g., plant ashes, potassium carbonate, or lye

Surfactants

Saponification

Hard soap is still made based on the same principle; it's just that now we use *lye* (sodium hydroxide, NaOH) instead of plant ashes or potassium salts. Lye is a stronger alkali and particularly suited to saponification. The reaction works with all possible fats; cheap fats like tallow, pig fat, and bone fat are traditionally used for hard soap. Might sound disgusting, but they make impeccable soap.

Why am I telling you all this? Because, given the modern trend for natural soaps, I find this traditional soap-making method to be particularly interesting. Natural soaps made from pure coconut oil, olive oil, or avocado oil are becoming increasingly popular. We are seeing a revival of the traditional soap recipe: pure fats saponified with the aid of lye. Natural soap is often nothing more than hard soap with attractive oils in place of pig tallow. Even natural soaps advertised as "100 percent saponified oil of whatever" are hard soaps by definition. Their chemical structure and properties are remarkably similar to those of hard soaps made from pig tallow. But natural soaps are frequently advertised as being particularly mild and caring for the skin. And yes, coconut, olive, and avocado sound totally mild and caring, but things look very different from a chemical perspective.

Hard soaps—both the "originals" and the "natural soaps"— are, above all, very effective. They offer very thorough cleaning because their hydrophilic group, the acid head, is so hydrophilic. However, this also means they're particularly aggressive. Obviously I don't mean aggressive in the same way as fluorine, but surfactants with lots of cleaning power can irritate or dry out the skin. And this is why we shouldn't shower every day. By giving ourselves a good clean, we intervene in the wonderful ecosystem of our skin flora, which also "cares for" our skin. And sebum isn't just produced to give us annoying pimples, but also to stop the skin from drying out. If the skin becomes too dry, it gets itchy and might even develop tiny cracks—and then the skin won't be able to fulfill its protective role, because bacteria and germs will get in.

You might think this is still much better than "chemistry." Natural soap fans have a particular hatred of one surfactant: *sodium laureth sulfate* (at least that's what it's normally called on the packaging; its actual name is sodium lauryl ether sulfate). If you don't like this surfactant, then you'll have a hard time at the drugstore, because this is by far the most common surfactant in shampoos and cosmetic products. It's a synthetic, "chemical" surfactant—and for some people, this alone is reason enough to replace it with pure olive oil soap. The "eth" in sodium laureth sulfate makes it slightly milder than hard soap, and may make it more suitable for cosmetics. In the surfactant pin structure, the "eth" (which stands for *ether*) can be visualized as a sort of connecting piece between the head and tail that sits somewhere in the middle of the hydrophilic–hydrophobic scale. The longer this connecting piece, the weaker the cleaning power and the milder the product is on the skin. So the fact that sodium laureth sulfate is a synthetic surfactant made in a lab doesn't automatically make it more aggressive—quite the opposite. Lab work produces various milder surfactants that can't be made through simple saponification and can be used in baby shampoos, for example.

I like natural soaps because they're environmentally friendly, but if you have sensitive or dry skin, you'd be better off using hard soap only to wash your hands. It saddens me when all synthetic surfactants are tarred with the same "chemical" brush. But the general distinction between natural and chemical soaps doesn't make total sense to me either. As I understand it, natural soap production is pure chemistry too. The avocado oil may be natural in origin (think of all the chemical advances aided by the avocado plant), but it can't be turned into soap without lye. And it is possible to produce environmentally friendly surfactants in laboratories. But products labeled as "chemical-free" sell better. You could call it "chemism," a form of discrimination against chemistry, the sweeping and totally unjustified attribution of negative traits. Down with chemism!

Whether we're talking about "natural" or "chemical" products, the real problem lies with shrewd cosmetics marketing. From a chemist's perspective, some things just don't add up. The best example is the new trend for marketing cleansing products with names like "micellar water," "micellar shampoo," and "micellar wipes." This "innovative micellar technology" is nothing more than a marketing ploy; by their very nature, every product that contains surfactants also contains micelles. Maybe someone should launch a "micellar toothpaste." I can see it now: "New! No fluoride! Featuring micellar technology!"

THE DOORBELL RINGS. I open the door to a sweaty, jubilant Matthias and feel myself overcome with a mixture of envy and guilt. Only yesterday, I received an email from my gym entitled "It's never too late!"

Reading my thoughts, Matthias grins and says, "Sitting is the new smoking!"

"Hmpf," I reply and sit down at my desk.

4

Sitting Is the New Smoking

ARE YOU SITTING comfortably? Maybe you should stand up for a while. Because...

"Sitting is the new smoking."

"The more you sit, the earlier you'll die."

"Lack of exercise kills twice as many people as smoking."

When I travel for work—which is often—I'm active most of the time, but when I work from home, I definitely don't move enough. Things are particularly bad at the moment because I'm putting a lot of time and effort into this book.

My usual work routine goes something like this: I get up at the same time as Matthias, roll out of bed, and go straight to my computer to "quickly check my emails." After what feels like an hour, Matthias walks back through the door and

I realize it's 6 p.m. I'm still wearing my pajamas and I've been working for eleven hours. All the more reason to make sure I get enough exercise. After all, it's "SCIENTIFICALLY PROVEN" that sitting is the new smoking. So go ahead, smoke all you want, as long as you're walking when you do it!

The general rule is that if something is "scientifically proven" and yet seems outlandish, then it's probably either not scientifically proven or not that outlandish after all. There are many truths to the statement that "sitting is the new smoking," but many exaggerations too. Let's start with the bad news—the facts.

Cardiovascular diseases, obesity, type 2 diabetes, cancer, and depression have all been linked to a "sedentary lifestyle" (the scientific term for "sitting a lot"). I'd love to know how long our ancestors spent sitting down. Did they perch on rocks or the ground at every opportunity? And was it only natural that the human race would eventually start to build chairs? Or does excessive sitting go against our biology; is it a (dangerous) cultural development? Experts tend toward the latter.

In his article "Sitting Is the New Smoking: Where Do We Stand?," physician Benjamin Baddeley writes:

> An alien visitor to our planet would be perplexed by modern human life, not least our relationship with physical exertion. After 6 million years of hunter-gatherer existence, humans can be observed sheltering in warm rooms, counteracting the tiresome effects of earth's gravity by slouching on comfortable seats in front of glowing screens, being whisked effortlessly between floors aboard mechanical staircases, even soaring across continents while seated in warm moving boxes.

Confusingly, however, a proportion of these same humans could later be spied spending their "free time" running around outside in all weathers for no apparent reason or, stranger still, handing over money to an institution called "the gym" to pass time repeatedly picking up and putting down heavy objects or running on a revolving mat until they were red and sweaty.

Yep. And I pay for the gym and don't even use it. I bet plenty of other inactive gym members are engaging in this special form of silent partnership.

As a species, we are succumbing to noncommunicable diseases (NCDs) with increasing frequency. These are diseases that can't be transmitted but still spread like modern epidemics, chronic diseases that advance slowly. The four main NCD categories are cardiovascular diseases (like heart attacks and strokes), cancer, chronic respiratory diseases, and type 2 diabetes. NCDs are responsible for 71 percent of all deaths worldwide; according to the World Health Organization (WHO), fifteen million people aged thirty to sixty-nine die each year from such diseases alone. I'm not saying this to ruin your day. I'm saying it because most NCDs are avoidable. We are responsible for the greatest risk factors: smoking, excessive alcohol consumption, an unhealthy diet—and lack of exercise.

We've long known that lack of exercise is bad and sport is good. And, obviously, that we don't move much when we're sitting down. But how dangerous is it really?

If you want to learn more, just search online for "sitting is the new smoking" and a whole bunch of articles will appear. Many of them say that even regular sport can't reverse the damage done by sitting (great, so it doesn't matter that I don't go to the gym—I spend so much time sitting that I'm basically doomed!). The important thing is to avoid long periods of sitting. Get up at least once an hour for a few minutes or, even better, use a standing desk—otherwise you'll undo all the health benefits of sweating in the gym or going for a run. Looking at it this way, sitting is more than just a lack of healthy movement; it actively damages your health.

So how true are these statements?

I will clear this up for you, but first we need to accept that in science there are often no short—and correct—answers. We like to assume that science produces clear facts, but that isn't necessarily the case. Science can produce clear numbers and measurements, but interpreting them is often so complex that we can't draw automatic conclusions. Sometimes we have a supposition and our experiments confirm that supposition. That doesn't make it a fact, merely a well-founded supposition.

We also like to assume that "scientifically proven" statements represent "the truth." But "the truth" is often nothing more than the current sum of all well-founded suppositions. New experiments may produce new findings that cause us to question everything we previously assumed to be "true." So if you want to think scientifically, you need to accept that there are no simple answers.

Here's an example. Say I invite some friends over for dinner. It's the first time that my work colleague—let's call him Paul—has come to my house. I don't know what Paul likes to eat or how much he normally eats. Anyhow, I make my signature risotto (which usually goes down really well) and, as

I expected, all the other guests eat it up with gusto. Paul is the only one who doesn't clear his plate, despite saying how much he's enjoying it.

Question: Why didn't Paul finish his meal?

There are various potential answers:

Paul didn't like it.

Or: Paul wasn't very hungry.

Or: Paul doesn't generally eat that much.

Or: Paul is on a diet.

Whatever, the answer can't be that complicated, can it? Well, if you asked a scientist, they'd explain it like this:

Paul ate less than everyone else at the dinner. It is possible that Paul either ate a lot less than the average or that the other guests ate a lot more than average. Paul was the only guest who didn't clear his plate. Other guests have consistently eaten larger portions of the same risotto in the past too. Together, these factors strongly suggest that the reason for this deviation in behavior lies with Paul himself.

There could be various reasons why Paul ate a smaller portion. It could be that he felt less hungry than the average—at this point, we cannot determine whether this is generally the case. It is possible, for example, that Paul ate more than usual at lunchtime, or that he doesn't usually eat much in the evening.

Another reason could be Paul's taste preferences. It is possible that he didn't like it. It may be that he doesn't like risotto in general, or that he didn't enjoy Mai's signature risotto in particular. This would contradict Paul's statements, which indicated that he enjoyed it. However, many people have been observed in the past to make statements like this that do not necessarily reflect the truth but are designed to strengthen social bonds with the host or follow other social etiquettes. As we are currently lacking information on Paul's other eating habits or social behaviors, this possibility would need to be considered with caution.

A combination of various factors is also possible, although we cannot currently quantify the influence of different factors.

Further investigations are required.

Are you still with me? Or has your brain switched off? If you've never read a scientific publication, this is EXACTLY what it's like. I'm not even exaggerating, I swear. Newspaper articles about science aren't usually this dry. Good science journalism is an art, of course, but sometimes articles interpret research findings incorrectly—or far too simplistically.

The problem is that it's very difficult to verify the actual research findings on the basis of a newspaper article. Obviously, you could get hold of the original study. But imagine that a quarter of the words in that passage about Paul's meal were specialist terms—that's the reality of scientific publications. They are unintelligible on multiple levels! They're written in a different language, a language for specialists, and this language is such detailed, nuanced gobbledygook that it's damned hard to identify the core message. I mean, the passage about Paul and his risotto meal is written in comprehensible language. But could you briefly summarize the links between Paul and the risotto? Not so easy, is it?

Clearly, it's very important for the media to communicate scientific findings; you wouldn't get far if you tried to read the studies themselves. Not to mention the fact that many studies aren't even available to the public—you have to pay to read them. This is why good science journalism is so important.

All journalists know that the public like simple answers. And simple headlines. Even better if they're both simple and dramatic! That's how we end up with headlines like "Sitting Is the New Smoking" or "Paul Hates Mai's Risotto."

The international media buzz around the dangers of sitting became so great that research was conducted into the media reports themselves. Australian communication scientists analyzed almost fifty newspaper articles (online and print) about the dangers of sitting. The results are very interesting:

First: Around a third of the articles stated that long periods of sitting are so dangerous that they counteract the health-enhancing effects of exercise. If this were true, sitting would certainly deserve to be described as "the new smoking." But is it true? In short, no! If you've just stood up in panic, you can sit back down. In reality, there is plenty of scientific evidence that sport and movement more than compensate for the negative consequences of sitting. If you sit a lot, then you're recommended to spend an hour to an hour and a half per day on your feet. But is there a difference between doing a little exercise every day and spending one day a week fulfilling my entire exercise quota? Even "weekend warriors" who only exercise one or two days a week can counteract the effects of extensive sitting ("only exercise one or two days"… the more I write, the guiltier I feel…).

It's dangerous to portray sitting as actively harming our health. Some people will read the newspapers, see that "sitting is the new smoking," and dig out their running shoes in fear. Those are the people who rush to the gym after Christmas or Thanksgiving to work off the cookies and turkey. But for me, that tends to kill my motivation; I'm one of the people who think, "Well, there's no point going to the gym now; what difference would it make?"

Looking at the science, sitting is definitely an underestimated risk. When thinking about exercise and sport, we should always factor long periods of sitting into the equation too. Science also confirms that sitting less is no easier than doing more exercise. And if we look at the "weekend warrior" study more closely, we can see that you would have to sit for several hours less per day to achieve an effect comparable to one or two days of sport per week. So if you're trying to decide between sitting less, going for a daily walk, or doing one intensive training session per week, you need to ask which option is most realistic for you.

It's the same with diets. The most effective diet is the one you can maintain. In my opinion, there's a far more constructive way of looking at all these scientific findings—"sitting less" has been added to the list of healthy exercises! This can be seen as an opportunity for people who don't want to or are not able to exercise. Of course, "Sitting Is the New Smoking" makes for a catchier headline than "Sitting Less Is the New Sport."

Second: A quarter of the articles emphasized that office workers are particularly at risk.

If you work in an office, then you're going to sit a lot. Obviously. If you look at the statistics, you'll see something very interesting: not all sitting is equal. Someone who sits a lot in their office has a healthier life than someone who sits in front of the TV a lot. Does that mean I can sit in the office without feeling guilty, but should stand up when watching TV?

This demonstrates perfectly that numbers alone do not produce meaningful facts. Allow me to put this in context: people who work in offices tend to have a high socio-economic status and a certain level of education, and they can afford a certain standard of living. In general, these factors promote better physical and mental health. Statistically speaking, wealthier people are also healthier.

Meanwhile, higher television consumption is statistically associated with a lower socioeconomic status, a lower level of education, and a higher rate of unemployment. These factors in turn correlate with poorer mental and physical health, such as an unhealthier diet. For example, people who watch a lot of TV are exposed to more ads for unhealthy food.

So you can see how complex it all becomes. It's relatively easy to find out how many people sit in front of the television or in an office. But it's practically impossible to use this data to work out how sitting influences a person's health. To put it another way, just because high television consumption is linked to a high risk of noncommunicable diseases, doesn't automatically mean that sitting is the dangerous factor in this complex equation.

There's one thing we can say for certain: noncommunicable diseases are strongly linked with socioeconomic status. On an international scale, 80 percent of NCDs affect people in lower-income countries. A typical pencil pusher is not the biggest victim, globally speaking. However, the media emphasis on sitting in offices is understandable because the articles and reports are aimed at people who work in offices. And perhaps they're even sitting in their office chairs as they read that they might as well be smoking.

Third: More than 90 percent of the articles conveyed the message that exercise is an individual's responsibility.

Isn't that true? I said previously that NCDs are avoidable, that we control the greatest risk factors ourselves. At this point, it's crucial that I explain this more precisely to avoid making the same mistakes as most articles on this topic.

Obviously, I am ultimately the one who forces my muscles to move, but there are so many external factors that influence my willpower. For example, I benefit from a level of prosperity that enables me to pay for a gym membership (in fact, I'm affluent enough that the membership fee alone isn't enough to motivate me to actually go). Plus, I'm self-employed; I don't have a boss or colleagues who'll give me the side-eye if I stand up every hour and do a few jumping jacks. And, as far as I can tell, I enjoy a level of mental health that enables me to maintain my physical health in the first place. And so on and so forth.

"Simply moving more" is easier said than done, but it's harder for some people than others. The fact that 80 percent of NCDs occur in lower-income countries has less to do with the willpower of the population and far more to do with their socioeconomic status and the many complex, interweaving issues.

We can combat NCDs in many ways, starting with education and continuing through to specific measures for employers to integrate into their workers' lives, from adjustable desks to established exercise breaks.

SO HOW CAN I apply all this to my everyday life? I won't feel so bad about my infrequent visits to the gym, at least until this book is finished. Instead, I'll actually get up every hour and do twenty jumping jacks, for example. Or go for a walk in the evening. Or both.

You're the one who decides what to take away from research findings. Above all, however, we must be prepared to eschew short answers and instead look at issues from as many perspectives as possible. To make good decisions, we need to understand an issue in detail.

So I'm going to take a break and get some exercise.

Want to join me?

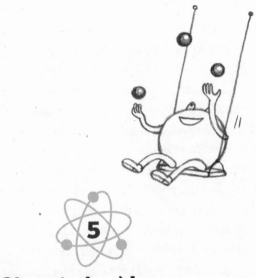

5

Chaotic by Nature

"AND WHAT ELSE do you do? Apart from YouTube?"

Most people assume that I must have another job as well as my YouTube channel. I do, but only because I'm crazy. Producing one science video a week would be more than enough for a full-time job, and I wouldn't get bored. Obviously I need to record and edit the video, but it takes just as much time—usually a lot more—to do the research and write the script. Before I had a fiber-optic connection installed in my apartment, it took me one whole day per week just to transfer the data; video files take up quite a few gigabytes. When your internet connection's that bad, it feels like it might be quicker to post the memory card.

I've just finished editing the video; all that remains is to upload it. My videos go online every Thursday at 6:30 a.m., but I can upload them earlier and set a release time. Sometimes I finish my videos only a few hours beforehand, in the middle of the night. That's just how it goes when YouTube isn't your only job. This time, I've managed to finish on Wednesday afternoon.

You can see my desk in some of my videos—clean and tidy, of course—but it's all a facade. My workspace often descends into shameful chaos. I'm not necessarily chaotic by nature; some aspects of my life are super organized, like my calendar and emails (basically everything I can arrange electronically). But in the analog world, it takes real effort to maintain some sort of order (with the exception of my wardrobe—I arrange my clothes by color). The chaos escalates on a regular basis, especially since I started working from home, where nobody is disturbed by my mess. Although I like to have visitors in theory, they're not allowed to drop by unannounced—they might catch me in one of my chaotic phases, which only my closest friends are allowed to see.

But why should we be ashamed of a bit of disorder? I see no logical reason to be embarrassed. Some people might say that a messy desk equals a messy life, but that doesn't make sense to me. If my chaos interfered with my work, then I'd pull myself together; there's not much I hate more than inefficiency. I always know where everything is. And as soon as I have to waste time searching, I know I need to clear up. In fact, I gain more time overall by doing less cleaning. It's logical really. I'm not chaotic, just pragmatic!

Admittedly, humans do seem to need a certain amount of order. There's some scientific evidence that order and behavior go hand in hand. Psychologist Katie Liljenquist has observed that the mere scent of citrus household cleaner can encourage more ethical behavior. In her experiment, two test groups were placed in two different rooms with identical furnishings. The only difference was that Room A was odorless, while Room B had been sprayed with a citrus-scented cleaner. Amazingly, the participants in Room B acted more fairly and generously during a staged game and also showed greater willingness to make charitable donations. Was this due to the association with cleaning, or did the citrus scent itself have a magical effect?

A few years later, psychologist Kathleen Vohs investigated the matter more closely, placing one test group in a tidy room and one in a disorderly room. Vohs then arranged for both groups to perform various unrelated tasks and fill out surveys, including a request to make a donation. As with the citrus-scented cleaner, morals and order seemed to be intertwined. The participants in the tidy room were willing to donate significantly more. At the end of the experiment, the participants were offered a choice between a sugary snack and an apple. The apple proved much more popular in the tidy room; the people in the messy room tended to go for the unhealthy snack. So it seems that humans need order and structure to conduct themselves properly.

I like talking about psychological studies; they often make for better stories than chemistry experiments. Unfortunately, I always find myself wondering whether these studies can be reproduced. *Reproducibility* means that if the same study is conducted again using the same methods, but with different participants, the same result will be achieved. This isn't always the case (or, more pessimistically, this isn't often the case).

A project published in 2015 saw 270 scientists join forces for one major test, in which they repeated ninety-eight psychological studies that had already been published. Less than half of the repeated studies returned the same result as the original studies—sobering, to say the least. But why? To be a bit more blunt, how the hell could this happen??

It's all a question of *scientific methods*—the way in which data is collected and analyzed. If you're interested in science, you need to remember that *scientific findings reveal very little if the methods aren't explained.*

Let's consider this in a bit more depth. Say you've developed a new medication and want to test its effectiveness in a clinical setting. The gold-standard method is a *randomized controlled trial* (RCT for short). What an unwieldy term. It's worth understanding what it means, particularly if you come across an online article that palms you off by simply citing the "results of a new study." Checking for an RCT will help you see these results for what they really are.

Let's take a moment to break down "randomized controlled trial." It should be more or less clear what a trial involves, but a closer look at the two adjectives shows why not all studies are equal.

Let's start with "controlled," returning to the fictional medication I just mentioned, the one you've developed and want to test. Let's make it a medication to combat procrastination, "postponitis," the unfortunate habit of putting off important tasks until it's almost too late (no such medication exists, but it would make the developer filthy rich). What do you do now?

Once the medication has been carefully tested in the laboratory, on cells and animals, it's time for the clinical trials. These involve giving your medication to as many people as possible and observing whether they become more productive and less likely to postpone things. That alone is not enough—a *control experiment* is essential, performed with a second group of subjects known as the control group. Instead of taking your medication, the control group is given a *placebo*, a pretend tablet with no effect. You can bet that the placebo effect will cause the control group to become more productive as well and to procrastinate less on average. Knowing (or believing) that you're taking a specific medication, and expecting a specific effect, often creates a self-fulfilling prophecy—I expect something to happen, so it does.

So your medication can only be considered effective if it causes the people in the test group to be significantly more productive than the control group people who received the placebos. Without this control, your study would be scientifically worthless.

I chose the example of anti-postponitis medication because it quickly shows how crucial the placebo effect would be. Motivation to be productive is psychological; we can persuade ourselves that it's happening (in Chapter 7, we'll explore why "persuasion" is fundamentally biological). But the placebo effect almost always plays a role in medicine, even with painkillers, allergy treatments, blood pressure medication, and so on. Really, we should carry placebo tablets around with us at all times to offer to our friends. If someone complains of a headache, simply get out the placebos and say, "Ooh, I just happen to have some headache tablets in my bag!" It works for other ailments too. "Here, these are for stomach pains," or "I've got these herbal relaxants; they always work."

On the other hand, there's the *nocebo effect*, the negative counterpart to placebos. Expecting undesirable side effects can also become a self-fulfilling prophecy. Participants regularly leave clinical trials due to side effects, unaware that they were actually in the placebo group, weren't taking any active substances, and therefore couldn't have had any side effects. This has been demonstrated with harmless injections of saline solution (I realize that "harmless injection" might seem like a contradiction). Although presenting no risk to people with food allergies, a placebo injection (containing no allergens) can trigger a genuine allergic reaction in some people.

Bearing the placebo and nocebo effects in mind, dividing trial participants into test and control groups isn't the only useful approach. It's equally important that the participants themselves don't know which group they are in—and that even the researchers performing and analyzing the trial don't know. When they look at a patient's data, they have no idea whether this particular participant has been given the real stuff or a placebo. Scientists are only human of course, and our personal expectations could affect how the experiment is analyzed and tarnish its objectivity, consciously or unconsciously. We don't trust ourselves, and that's a good thing. This method is known as *blinding*, and a study in which the participants and scientists are all kept in the dark is called a *double-blind study*. The blinding may only be lifted once the data have been analyzed.

So a high-quality clinical trial must be "controlled"—and ideally "double-blind." Now let's consider "randomized." Randomized is not the same as random. It means that randomness is deliberately forced. But how?

Let's go back to our anti-procrastination medication. If we want to test whether it works, then obviously we want it to work. This means there's a risk that—consciously or unconsciously—we will assign participants to the test and control groups in whatever way most benefits our goals. I could put the participants who are already more productive and procrastinate less into my test group and place the less productive people into the placebo group. By doing so, I would falsify the results. To prevent this from happening, the trial participants are randomized—a computer program randomly assigns them to one of the two groups so that I, the scientist, can't exert any influence.

HOPEFULLY YOU NOW understand why an RCT is the gold standard in clinical trials. Despite every care being taken, medicine is so multilayered and complex that, even with the best methods, simple answers don't come automatically (as we saw in Chapter 4). Even in the best randomized controlled trials, it may not be possible to reproduce a result.

Poor reproducibility is a particular problem in psychology. This doesn't necessarily mean that psychologists are doing bad work, rather that psychological methods aren't as watertight as RCTs. For example, psychological studies are often based on surveys, so on participants' own statements. How far can these be trusted?

I'm sure you can answer that one for yourselves. Unfortunately, there's no better way to find out how somebody feels, other than to ask them. But even extensive conversations with experts, who can evaluate the participants, are also susceptible to error because a researcher's qualitative analysis is not the same as a physical measurement that produces a numerical value. Obviously, experiments like these are harder to reproduce. Also, statistics don't mean anything unless an experiment has a suitable number of participants. A good trial examines as many participants as possible. So researchers need to find methods that can be realistically implemented (and reproduced) on a large scale. This allows scientists to work conscientiously and correctly and still make mistakes—errors are central to their methods. This is exactly what I meant before: when you're looking at scientific results, always pay close attention to the methods. Ask yourself how exactly the results were achieved. Results alone can be misleading. I'm not trying to say that all psychological studies are trash. You can enjoy results like "citrus cleaning products make us well-behaved" as long as you remember to view them critically. An appropriate response would be "How interesting!" not "I need to spray my children's bedrooms immediately!"

Unfortunately, many people misunderstand science and are too quick to apply scientific findings to practical situations. The idea that order encourages us to act morally is reflected in the *broken windows theory*, used in crime prevention. According to this theory, minor violations like trash on the street, graffiti on walls, and broken windows themselves must be strictly prevented because harmless offenses lead to serious crimes. They must be tackled rigorously and punished severely because tidiness and cleanliness prevent criminality. This remains a controversial approach—the punishments are excessive and there's no proof that they work—and yet some people still reference the scientific studies that I previously mentioned. Again, this clearly shows that scientific findings are often nothing more than well-founded suppositions, and not always suited to direct application.

ACCORDINGLY, KATHLEEN VOHS and her colleagues asked themselves why chaos always prevails if humans have such a need for structure and order. Maybe we need chaos just as much as order? Again, Vohs used two trial rooms (one tidy and one untidy) to test a supposition that, while popular, has little scientific evidence so far: chaos equals creativity.

Here's a creative task for you: Imagine that you run a factory that makes table tennis balls. The number of people playing table tennis is steadily declining. To stop your company from going under, you need to find a new use for your table tennis balls. How many different ideas can you come up with? Be as innovative as you like; wild ideas are welcome, even if they would be difficult to implement.

Vohs presented this same task to both groups of participants. The group in the untidy room had more creative and unconventional ideas for repurposing table tennis balls, for example as ice cube trays and in models of molecular structures. Unconventional thinking is one of the core skills of creative people, the ability to see things in an unusual context, to make connections between unrelated things. Disorder seems to help with this.

In another experiment, Vohs divided participants into two rooms and offered them fruit smoothies labeled either "classic" or "new." The participants in the tidy room preferred the "classic" drink, while those in the untidy room tended to go for the "new" version. So chaos encourages us to try the unfamiliar, the unconventional, the new. Vohs identified a positive side to chaos. Without creative and innovative thinking, without the courage to try new things, we would have no art, no scientific progress. And despite everything I've said about approaching psychological studies with a critical eye, I admit that I like to quote these specific findings, particularly when I have unexpected visitors and expose them to my chaos.

AS A CHEMIST, I like to consider chaos from a thermodynamic perspective too. *Thermodynamics* is wonderful, bringing physics and chemistry together. The laws of thermodynamics are a bit like human rights, only for molecules. They don't distinguish between different particles. It doesn't matter whether you're an oxygen molecule or a gold atom; the laws of thermodynamics apply to us all—all living beings, all objects, all molecules, all atoms, all physical and chemical processes.

Thermodynamics and quantum mechanics comprise the most fundamental scientific understanding of this world and universe. And thermodynamics says that *the universe doesn't just want to be chaotic, it has to be chaotic.* Otherwise, I might suddenly suffocate while writing this sentence, because all the air molecules in my room have suddenly gathered in one corner and left me with nothing to breathe. You might think that sounds absurd, but is it really?

The air around us is 78 percent nitrogen gas and 21 percent oxygen gas; the remaining 1 percent is made up of noble gases and carbon dioxide. Together, these gas molecules make up less than 0.1 percent of the volume of air, of the space. The rest is nothing!

Now, look at the tip of your pinky finger. Obviously we all have differently sized fingers, but let's say that the tip of your pinky has a volume of one cubic centimeter (less than one-tenth of a cubic inch). An air volume of one cubic centimeter contains around twenty-six trillion gas molecules. That's twenty digits before the decimal point.

Naturally, these air molecules also have a mass. A single molecule doesn't weigh much, but each cubic meter (35 cubic feet) totals around 1.2 kilograms (2.6 pounds). This is known as *air density*.

This incredible swarm of molecules doesn't just have a mass; it's moving too! The speed at which the molecules move depends on the temperature. Remember our cup of coffee in Chapter 1? The warmer, the faster. At room temperature, gas molecules rush around us at a speed of over 1,000 km/h (620 mph). These little molecules are pretty boisterous, and they exert pressure too. Pressure is the force per unit area. By constantly colliding with us and every other surface, the molecules exert pressure—*air pressure*. This is around 1 bar, which equates to a weight of 10,000 kilograms per square meter (2,000 pounds per square foot).

Let's assume that the area above my head is 0.1 square meters (1 square foot). That means I am subject to air pressure of around 1,000 kilograms or 2,200 pounds—1 metric ton—almost as much as a vw Golf (the most commonly sold car in Germany). The same goes for you; we are all being constantly bombarded by air molecules. How do we stand it? Why don't we feel this massive air pressure?

Well, we're made from molecules too. And the molecules inside us exert outward pressure equal to the atmospheric pressure. If the pressure outside us changes, our eardrums usually let us know pretty quickly. The thin membrane in our ears remains unnoticed as long as the pressure on both sides of the eardrum remains the same. As soon as the outside air pressure changes—for example, when an airplane takes off or lands—it feels as though our ears are closing up. When the outside pressure increases, air molecules pound against the ear from the outside and press the eardrum inward; if the outside pressure decreases, air molecules pound against the ear from the inside and press the eardrum outward. This is what makes our ears feel like they're blocked. And now that the eardrum has less freedom to vibrate, our hearing is muffled. However, the ear has a kind of valve, the Eustachian tube, which connects the ear and the nasopharyngeal space. It's normally closed but opens briefly when we chew or yawn, which evens out the pressure.

On an airplane, we feel the difference in pressure because the air density decreases the higher we go. What would happen if we went even higher—into space, for example?

Earth's thick atmosphere full of wild air molecules is rare in our universe. The vastness of space is a *vacuum*, a term indicating empty space, no molecules, nothing. What would happen if a human were launched into the vacuum of space without a protective suit? Spoiler: they'd die. How exactly? That's the interesting question.

This scenario has been explored in lots of science fiction movies. In *Star Wars: Episode* VIII, for example, Leia is hurled into space and seems at first to be freezing; her skin looks to be covered in tiny ice crystals. Some *Star Wars* fans have criticized this scene for being too unrealistic—in the end, Leia survives by using the Force to pull herself back to the ship. I mainly find this scene unrealistic because you wouldn't freeze that quickly in space, even though space is unbelievably cold. By "unbelievably cold," I mean close to *absolute zero*, the coldest possible temperature.

If temperature is nothing more than the movement of particles (as explained in Chapter 1), then "as cold as possible" means "as slow as possible." Absolute zero, 0 degrees Kelvin or –273.15°C (–459.67°F) can therefore be envisaged as the point at which movement stops completely.

Things can't get any colder than this, which is why there is a physical lower limit for temperatures. In practice, however, the *third law of thermodynamics* prohibits absolute zero from actually being reached. Space comes close, at 2.7 degrees Kelvin or –270.45°C (–454.81°F). How would you *not* freeze in that?

Again, this is connected to our coffee in Chapter 1. Cooling works mainly through heat conduction, which requires molecules to collide. The more contact between matter—the more often particles can collide—the better heat conduction works. This is why, for example, a bucket of icy water cools down drinks much more efficiently than a bucket of ice cubes (just a little tip in case you have unexpected guests and need to chill some drinks). There is air between the ice cubes, and air is made up of fewer particles than water, which means fewer particle collisions. Bottles cool slowest in fridges; air is a lousy heat conductor.

The vacuum of space is even worse. Space has no matter at all, meaning there are no molecules to which a warm body can transfer its heat. We would cool down through heat radiation, but only very slowly. Despite temperatures close to absolute zero, we wouldn't freeze that quickly in space.

VACUUM

But might we explode? Would we burst if there were no external air pressure to counteract the pressure inside our bodies? There are videos on YouTube showing Twinkies placed in vacuum bell jars from which the air is slowly extracted. The Twinkies inflate and the white filling bursts through the cake layer. Luckily, we are not Twinkies. Our skin and tissue are robust enough to hold us together in a vacuum.

We don't need to explode. We could die less dramatically, and still highly unpleasantly. Even in the upper reaches of the Earth's atmosphere, from around 18–19 kilometers (11–12 miles) above sea level, our bodies would kind of start to bubble. This is called ebullism (from the Latin *ebullire*, meaning "to bubble out"), and the symptoms are a lot less fun than "bubbly" makes it sound: the water in the mouth and eyes begins to boil, blood circulation and breathing are impaired, the brain doesn't get as much oxygen as it needs because the arteries are blocked, and the lungs may swell and hemorrhage. A vw Golf full of air molecules on your head is far more preferable.

Why would our bodies begin to bubble? Bubbling means boiling, a liquid converted to a gas. There are basically two ways to make something boil. The first is to heat it. We talked about this in Chapter 1: when I boil water, I make the water molecules move more. At some point, the need to move becomes so great that the water molecules are no longer willing or able to keep hold of one another—so they evaporate. The other way to boil liquids is to reduce the pressure. To be more precise, water only begins to boil at 100°C (212°F) if the atmospheric pressure is the same as at sea level. At the top of Mount Everest, at 8,848 meters

(29,029 feet), decreased air pressure means that water boils at 70°C (159°F). Water is liquid because its molecules keep hold of one another, but also because of air pressure. The air molecules collide with and press against the water molecules in my cooking pot, just as they do to us. If I took my cooking pot to the top of Mount Everest, the air would be thinner and there would be fewer air molecules to collide with the water. The lower the air pressure, the easier it is for the water molecules to leave the pot and evaporate. And at a height of 18–19 kilometers (11–12 miles), the air is so thin that water evaporates at body temperature. In space, where there is practically no air pressure, evaporation is as easy as it gets.

The air in our lungs would also expand a lot. Just as there are two ways to change states of matter, there are two ways to influence the volume of a gas. The first is temperature: *the lower the temperature, the less space a gas takes up; the higher the temperature, the more space it takes up.* To see this for yourself, simply take a bottle and replace the cap with a balloon. Warm up the bottle, for example in a hot bath, and the balloon will "blow itself up" (provided there's no way for air to escape the bottle). If you cool the bottle down again in cold water, the balloon will collapse.

The second way is through pressure: if I blow up a balloon, the air molecules inside the balloon collide with the skin of the balloon. The size of the balloon depends not just on how much air I blow in, but also on the air pressure from outside. If I now reduce the pressure from outside, the balloon would expand further. So a balloon made up almost entirely of gas would, in fact, explode in space.

HOME EXPERIMENT NO. 2

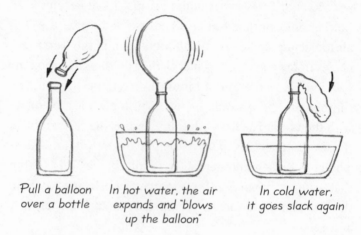

Pull a balloon over a bottle

In hot water, the air expands and "blows up the balloon"

In cold water, it goes slack again

If I were hurled into space, my instinct would probably be to hold my breath—which would quickly expand and burst my lungs. It would be better to breathe out quickly (but I'd die anyway, so it wouldn't actually matter).

Even if I didn't explode like a piece of candy and my skin and tissues stayed intact, I wouldn't survive for long. In the 1960s, astronaut Jim LeBlanc narrowly escaped with his life after his space suit depressurized during an exercise in a vacuum chamber. The last thing he remembered before losing consciousness was a tickle in his mouth from his bubbling saliva. Luckily, he was returned to the thick air in good time and suffered no lasting damage.

So things will start bubbling, but we'd probably die from lack of oxygen first—a totally unspectacular death. Luckily, we'd lose consciousness after a few seconds without oxygen, so we wouldn't be aware of all we suffered. And this is why

I agree with the *Star Wars* critics; it's extremely unrealistic for Leia to be able to use the Force (which takes a lot of mental power) in the void of space.

Here's another interesting question: What would happen to a dead body in space? Basically, not a lot. The usual decomposition processes that take place on Earth are chemical reactions with oxygen and water, which require a bit of warmth and sometimes the help of microorganisms. Space has none of these things, so a corpse would be perfectly preserved. Well, I suppose that's something.

LET'S COME BACK down to Earth before things get too macabre. Thankfully, I have plenty of air molecules whizzing around me as I sit at my desk and upload my video. The door and window are closed, so they can't get out, but they can move freely within the office. Nobody tells air molecules which direction to move in, so it is theoretically possible that all the molecules could end up in one corner of the room, there'd be no air at my desk, and I would suffer a space death on Earth. It's definitely possible, but very, very improbable. This is why people working in physical chemistry very rarely talk about the "possible" and "impossible," sticking to *probabilities* instead. This is how you can spot physical chemists in day-to-day conversation. Instead of saying, "I could never run a marathon," they say, "It's extremely improbable that I would manage a marathon."

This improbability is based on our old friend, the second law of thermodynamics. In Chapter 1, I said, "Close the door, the heat's getting out!" Heat always flows from warm to cold, never the other way around. There's a much more general way to express this: *chaos never abates by itself.* This affects both the universe as a whole and my desk specifically. My desk will never become more orderly by itself, only more chaotic. Every spontaneous process in this universe, be it physical, chemical, or biological, leads to greater chaos than before—unless effort and energy are spent to reverse or stop the process (like me cleaning my desk). In fact, chaos has its own specialist term: *entropy. The entropy in the universe is constantly growing.*

So how does this help me stay alive at my desk? If all the gas molecules in my room move freely and independently and fly around as randomly and chaotically as possible, then the logical consequence is a fairly even distribution through-out the room. If you wanted to gather all the molecules in one corner of the room, you would need to restrict their free movement and bring order to the system. And that goes against the law of the universe. Which means it's impossible—er, I mean improbable, of course.

The second law of thermodynamics also ensures that everything in this room is at room temperature (with the exception of my body, which works to stay at 37°C [around 99°F]). If I bring a hot coffee into the room, it would go against the law of chaos for the coffee particles not to travel throughout the room. In the end, chaos leads to the most even distribution possible. If I wanted my clothes to be as chaotic as possible, I wouldn't arrange them by color in my wardrobe; instead, I would throw them all over the room, ultimately leading to an even distribution.

I take a deep breath of air molecules and stare at my screen, feeling tense. My video isn't uploading. With a sense of foreboding, I go into the living room and stare in dismay at the Wi-Fi router. The lights have gone out.

What's in It for Me?

A FEW WEEKS ago, I was on a train, in a lousy mood. I went to the onboard restaurant and bought a Belgian waffle with cherries and vanilla ice cream, one of the most popular dishes on the ICE trains that run across Germany. I had intended to eat it in the restaurant but was joined by two besuited men. The older of the two, who could have been my father, said, "Ooh, that looks good!" After a brief pause, in which he leaned a little closer, he followed it up with, "And the waffle doesn't look bad either—hahaha!" His younger colleague laughed along sheepishly.

The "slow blink" is a great way to express silent disdain: keeping your face entirely free of expression, simply raise your eyebrows, maintain eye contact, and blink once very slowly, for around a second. So I looked up from my Belgian waffle and raised my eyebrows. First I stared at the younger man, who I was pleased to see practically oozing discomfort, then I turned to the older man, who also seemed somewhat unsettled by my silent eye contact, and then I blinked nice and slowly. Then my good-looking Belgian waffle and I stood up without a word and left the two idiots sitting in the restaurant.

In the past, I would have either laughed along nervously or raised my voice and verbally abused them. But since I started making YouTube videos and receiving thousands of comments every week, I've been following the noble gas configuration method. Let me explain—it might help you too.

Noble gases can be found in the eighth main group of the periodic table, and they provide great inspiration for interpersonal tensions. The best-known noble gases are the first two in the group; we know helium (He) from balloons and neon (Ne) from tube lights. If you continue down through the group, you'll find argon (Ar), krypton (Kr), xenon (Xe), radon (Rn), and oganesson (Og). Think back to Chapter 2—the elements in the periodic table are organized by atomic number, the number of protons in the nucleus. So the farther down you go through a group in the periodic table, the heavier the elements get. The heavy nuclei are often radioactive, which means they are unstable and decay. Of the noble gases, radon and oganesson are radioactive, so we'll leave them aside for now; the other noble gases are known for their remarkable stability. Noble gases are unbelievably inert—frankly, they can't be bothered to bond with just anyone. And this is what makes them so "noble."

Let's compare noble gases with the fluorine from Chapter 2. The whole purpose of fluorine's short, aggressive life is to find a binding partner. This just isn't necessary for noble gases because they have eight valence electrons (the octet rule again). Fluorine won't find peace until it's joined with carbon to make Teflon or with sodium in toothpaste, but noble gases are perfectly happy from the get-go. While the other elements in the main groups form bonds to get to eight valence electrons, a noble gas has everything it could desire. This is why the octet rule is sometimes also known as the *noble gas rule*, and a full outer shell is called the *noble gas configuration*—the name I also use for a relaxed, positive attitude. It's a great mantra to repeat when you find yourself irritated by stupid trivialities. "Be a noble gas!" I tell myself.

Noble gases are so self-satisfied that they don't even form bonds with each other. For comparison, other gases like nitrogen gas and oxygen gas are built from *dimers*. A nitrogen molecule is made up of two nitrogen atoms, N_2, and an oxygen molecule is made up of two oxygen atoms, O_2. The same goes for hydrogen gas, H_2. But noble gases are content to float around by themselves; they are *monomers*.

Chemically speaking, noble gases aren't particularly interesting because they never have anything going on. Quite the opposite—the noble gas argon is used as an *inert gas* in laboratories (along with nitrogen gas). Imagine that you want to perform a chemical reaction in your lab but the substances you're using are so sensitive to air that they would immediately react with the oxygen and the whole experiment would be a bust. Other substances are extremely sensitive to water, and even the air moisture in a normal room would be enough to ruin them.

Well, I mean, I'm saying they'd be ruined, but a substance that's supersensitive to oxygen just really loves reacting with oxygen. Chemists have the ability to create forced marriages, to combine substances that would never come together otherwise. One way to do this is to force all the air out of the flask containing the substance and place the entire apparatus in an argon atmosphere. Once the inert argon has replaced all the air, you can perform your reactions without oxygen or air moisture getting in the way.

If noble gases weren't so inert, inhaling helium to make your voice go funny would be a very bad idea. Pure oxygen gas, for example, should only be enjoyed with caution (we'll come back to this in Chapter 10).

Noble gases do their own thing and rise above it all. They're very inspiring when you have to deal with disgusting older men on trains. With the noble gas configuration, you don't even need to respond. My Belgian waffle and I can simply glide away like a monomer argon atom.

BACK AT MY seat, however, my calm and noble attitude was ruined when I realized that the onboard Wi-Fi wasn't working. (This is what people around the world don't know about Germany, the country of engineers: our Wi-Fi sucks.) Then my bad mood returned. I had loads of work to do. Another older—but more pleasant—gentleman sitting opposite me smiled wisely. "These days we're lost without the internet, aren't we?" he said, offering me his newspaper. I didn't bother explaining to him what "the internet" means these days. Anyway, now I'm standing in the corner of my balcony, squashed against the railing, twisting my torso to get a phone signal. (Yes, this is Germany. Our phone signals suck too.) I'm cut off from the outside world unless I stand in this specific corner of my balcony, where I at least get enough signal to call Christine.

"Can I stop by? My internet's failed and I need to upload a video."

"What, again?" It's the second time in a month that my internet has failed, which is particularly disastrous in an apartment with no cell reception. I should really get insurance. "Sure, come round; I could use a distraction."

I'm so thankful that Christine and I have ended up in the same city again. We met during our PhDs—we had the same supervisor—and supported each other as we pursued our dreams. You know, the fact that we have PhDs doesn't mean we're especially clever. Around 85 percent of chemistry students follow their master's degree with a doctorate. It's practically a standard part of chemistry training. The main thing you need to do a doctorate is a high frustration threshold; be proud that you've passed the finish line, but don't get too conceited.

REGARDLESS OF HER academic pedigree, Christine really is very smart. After her doctorate, she did a postdoc in the United States. Postdocs are similar to PhD students; they already have a doctorate, but allow themselves to continue being exploited by universities. My cynical description of a university's academic hierarchy would be that the professors sit at the top like the gods on Mount Olympus, while the PhD students are at the bottom, doing the grunt work for peanuts, and the postdocs are trying to fight their way out of the masses; pre-graduate students aren't even worth mentioning. On the academic career path, in many countries a postdoc leads to a "habilitation" (a postdoctoral qualification required to become a professor) or a junior professorship; now, however, more and more businesses (such as pharmaceutical companies) require postdoc experience for entry-level positions. It's an absurd inflation of qualifications.

Meanwhile, my friend Daniel, who couldn't even derive an exponential function in high school, gained a bachelor's degree in business and earns more than any postdoc. It isn't necessarily the salary that scares people away from academic careers. To progress at a university, you need to sacrifice your private life and sleep on the altar of science. Christine never goes a weekend without working. Although we live in the same city, I usually only see her when I visit her lab. And all this work doesn't guarantee job security. You move from one fixed-term contract to the next. And an academic career path has just one destination: a tenured professorship. If you're extremely good, you might reach this point by your late thirties. Although very few people—and only the very best—make it to the habilitation stage, there simply aren't enough tenured professorships. Hard work, intelligence, and talent aren't enough; you also need to be really lucky. If you don't manage to secure tenure, then at some point you'll find yourself totally overqualified, possibly even applying for the same industry jobs as your own students.

Why would anyone do that to themselves? All for science? Research requires a healthy dose of idealism and a strong belief in the basic principles of science—a desire to pursue a career that will benefit society. I know it's trite, but they want to serve humanity and make the world a better place! This mindset does have its dangers though, as shown by Ranga Yogeshwar, an influential German science journalist who wrote the opinion piece "Und was bringt's mir?" (What's in it for me?). In the article, he reflects on the "value" of science and warns against the commercialization of research, or to put it more simply, against asking, "What's it worth?" Here's an extract:

It is a mistake to [...] reduce scientific curiosity, and the deep-seated urge to better understand the world, to economic categories. Not everything that is researched fits into this approach. For example, what is the "worth" of Collaborative Research Centre 933, which develops new approaches to interpreting ancient and medieval texts? It receives funding of 11.5 million euros. The DFG [German Research Foundation] also supports a project focused on the birthing center of the ancient Egyptian temple of Edfu; our taxes are being used to translate the hieroglyphics in its sanctuary, sacrificial hall, and large parts of its vestibule. What is the value of this research? Translating ancient Egyptian hieroglyphics probably won't increase gross national product or help the economy to flourish. But isn't this still great science? It strives to solve the puzzle of an ancient civilization and expands our understanding of the past. There is no economic "return on investment" in researching the properties of the Higgs-Boson or proving gravitational waves, and closer inspection reveals that even the economic spin-offs in these disciplines are not good arguments. For

*decades, I have watched a scientific landscape that strives to
legitimize its inner driving forces—curiosity and knowledge—to
the outside world using utilitarian arguments. [...]*

*Isn't it time for science to present a counterpoint to
one-dimensional economic perspectives with confidence and
passion?*

Initially, it's gratifying to think that research aims to serve
humanity. And yet our interests, as humans, often tend to
be economic and financial. This focus on benefits jeopar-
dizes the independence of research. And this independence
is essential in our society—a voice that speaks the truth and
goes beyond selfish or monetary interests. Only then can science truly make the world a better place.

Even scientists ask, "What's in it for me?" Yogeshwar
continues:

*At Google DeepMind in London, for example, personnel costs
for its four hundred employees amounted to 138 million dollars
in 2016, an average of 345,000 US dollars per employee. The
consequence of this is that outstanding scientists leave public
research institutions and universities in order to serve large,
well-paying private companies. "What's in it for me?" drains
independent, public institutions of their expertise, leaving them
with fewer and fewer great minds.*

If these "great minds" translate "What's in it for me?" into
"I won't let myself be exploited," or "I'd like a private life
too," or "I have some self-respect," then it's no wonder that
more and more excellent people are turning their back on

academia and moving to industry, where they'll at least earn a decent salary.

Christine has decided that her thirty-fifth birthday will be her cutoff point. If she still sees no serious prospect of a tenured professorship, then she's out. She's the youngest junior professor at the research institute and has been very successful so far. Only time will tell if she can gain a long-term foothold in the sciences.

She did her postdoc at the Massachusetts Institute of Technology (MIT) in Boston, every nerd's dream. During that time, she was courted by a whole range of companies. She was flown to Germany for interviews in the chemicals industry and was offered a job at McKinsey. But she wanted to stay in research, ultimately accepting her current position as a junior professor. I find that both stupid and honorable—stupid because she's exploited by the system, and honorable because we need people like her to shake up the system by resisting the question of "What's in it for me?" And, of course, I'm delighted that she's back in Germany and in my neighborhood.

Christine is my closest connection to the world of laboratories and universities. Sometimes, she lets me film sketches in her lab. And she has a fast internet connection, which has saved my ass before. I copy my video onto a USB stick and head to the institute.

The institute is an interdisciplinary research center that aims to break down the barriers between disciplines; Christine works closely with an engineer and a computer scientist. These days, you won't get far unless you engage with people outside your subject area. The major problems of the modern world are interdisciplinary in nature and require interdisciplinary solutions. Besides, the boundaries between disciplines are pretty arbitrary. Christine's research is much closer to physics than to some areas of chemistry. During our doctorates, our respective areas of chemistry research were so different that we never managed to get a shared project off the ground, however much we wanted to.

CHEMISTRY CAN BASICALLY be divided into three main areas: *inorganic chemistry, organic chemistry,* and *physical chemistry.* There are others, of course, but these are the three basic areas taught in depth in all chemistry study programs.

As the name suggests, *physical chemistry* is the link between physics and chemistry. It includes thermodynamics and quantum mechanics, but also Christine's field of expertise—predicting chemical reactions with the aid of computer simulations.

Organic chemistry mainly revolves around a single element—*carbon* (C)—and everything that likes to bond with carbon, including hydrogen (H), oxygen (O), nitrogen (N), and phosphorus (P). So organic chemistry focuses "solely" on all compounds and reactions containing carbon. I've put "solely" in quotation marks because organic chemistry is a huge area. All life on Earth is carbon-based. We are built from carbon. This book is built from carbon. Every structural chemical formula you'll see in this book is based on carbon. We can even assume that extraterrestrial life—if it exists—must be based on carbon too; no other element has such a diverse chemistry.

If you're reading this book on an e-reader or tablet, then *inorganic chemistry* also plays a major role. Inorganic, literally "non-organic," concentrates on everything that isn't carbon. This might sound like a lot when you look at the periodic table, but it doesn't initially seem like that much in nature—mainly salts, minerals, and metals. Some organic chemists claim, disparagingly, that inorganic chemists spend their lives studying rocks. First of all, rocks aren't actually that boring. And second, inorganic chemistry can be much more than that, and it can be pretty cool. It's the basis for all technological gadgets; smartphones are a masterpiece of inorganic chemistry. And with a little understanding, you can find out (among other things) how to make your cell battery last longer—that alone is worth a closer look at the chemistry in a phone, right?

Our smartphones are made up of over seventy different elements. Carbon is one of them, but the metals are what make phones so interesting; the microelectronics contain a few hundred milligrams (seven one-thousandths of an ounce) of silver and around 30 milligrams (one one-thousandth of an ounce) of gold. The screen is permeated by a very fine network of indium tin oxide fibers, which pick up on the electrical conductivity of our fingers and add the "touch" to the screen.

What really makes a smartphone "smart" is a special group of noble metals known as *rare earths* or *rare earth metals*. They can be found in the subgroups of the periodic table and include scandium (Sc), yttrium (Y), and the whole series known collectively as *lanthanides*. The lanthanides are usually shown below the periodic table with the actinides; otherwise the table would be far too broad.

The rare earths have some pretty spectacular names. Terbium (Tb), praseodymium (Pr), yttrium (Y), gadolinium (Gd), and europium (Eu) provide the vibrant display colors that make our Instagram pictures look so good. Neodymium (Nd) and dysprosium (Dy) turn normal magnets into the super-magnets used in speakers and microphones. These two rare earths are also used in the technologies that give our phones a "vibrate" mode.

Rare earths give energy-saving lightbulbs a more natural light and are used in green technologies like solar cells and wind power stations. They take effect even in tiny quantities, earning them the German nickname "Gewürzmetalle," which translates to "spice metals" (an insult to sodium, a metal in cooking salt that has surely earned the right to be a "spice metal"!).

Although rare earths are sometimes used in sustainable technologies, their extraction is anything but. These metals are not as rare as their name might suggest. They're very common in the Earth's crust, but are often dispersed throughout the stone. This makes extraction expensive and energy-intensive, and the working conditions are often unfair. China has the largest deposits of rare earth metals; it practically has a monopoly on some of these raw materials. This is a major competitive advantage in a high-tech age in which I have to leave my apartment as soon as the Wi-Fi fails.

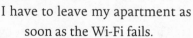

The more phones we buy, the more expensive rare earth metals become. And yet phone providers entice us with contracts offering yearly smartphone upgrades. If our devices aren't used more sustainably and recycled more wisely, then we'll soon have a problem. Plastic waste isn't the only sustainability issue in our modern society. Of course, most phones and tablets are deliberately constructed in a way that's difficult to repair. Many people treat themselves to a new device when the battery reaches the end of its life or the screen gets broken (we could send it away for repair, but how would we manage without our smartphones for more than a day?).

Incidentally, the fact that phone screens break so easily has nothing to do with manufacturers' bad intentions. The screens are actually remarkably robust and a little stroke of chemical genius. Instead of conventional glass, most devices use impressive-sounding Gorilla Glass®. This is made with *aluminosilicate glass*, comprising silicon, aluminum, and oxygen atoms, which together form a three-dimensional framework with a negative charge. This is balanced out by positively charged sodium ions, which sit in the gaps in the framework. To make Gorilla Glass, the aluminosilicate glass must be immersed in a hot saline solution containing positive *potassium* ions.

We know potassium from the potassium salts used in soap production. If you look at the periodic table, you'll see potassium (K) directly below sodium (Na). Potassium is another *alkali metal*. Just like sodium, potassium prefers to be a positively charged ion (the octet rule), but the potassium ion is a great deal bigger than the sodium ion.

So what does this mean for Gorilla Glass? When aluminosilicate glass is immersed in a hot potassium saline solution, the larger potassium ions force the sodium ions out of the spaces in the silicon framework and squash themselves in. This uncomfortable-sounding swap is made easier by the high temperature and the fact that all the particles are moving more overall. When the glass is cooled down again, it has a slightly altered chemical structure, with slightly too-fat potassium ions in slightly too-small spaces. This ensures greater compression and a more robust material: Gorilla Glass.

So why do so many phone screens break? Well, normal glass would break even more often. Whether a phone screen survives a fall depends on how strongly the glass is shaken. If the phone is relatively flat when it hits the ground, there's a good chance that not much will happen; the impact is spread over a larger area. Having discussed air pressure in Chapter 5, we know that pressure is force per area, and the smaller the impact area, the greater the pressure. If the phone lands on just one small corner of the screen, there'll be lots of impact, and you'd have to be very lucky not to break the screen.

By the same logic, if you ever find yourself in a plummeting elevator, you should try to lie flat on the floor, spreading the impact over a larger area. But once you're falling, almost in a state of artificial weightlessness, it's tricky to lie down. To be totally safe, you'd have to lie down as soon as you entered the elevator, which might seem strange to the other people in there. That's just a side note; elevators are well secured and won't fall.

Anyway, the best thing to do is to get a protective phone case, even if it makes your beautiful smartphone look bulky. I've never broken my screen. But as I take the bus to Christine's lab, I realize that my phone battery's almost empty, so I switch it off.

Battery life is so annoying, isn't it? Do you know how long cell or mobile phones could go without charging fifteen years ago? Three to six days, depending on how much *Snake* you played. These days I'm glad if I can go a whole day without charging my phone. Once you understand the chemistry of a phone battery, you'll also understand how best to treat your battery to make it last longer. And this is the other side of the "What's in it for me?" question. Again and again, I hear people say that we don't need science later in life, so there's no point paying attention in school. Hopefully by now I've convinced you otherwise.

So: there are different types of rechargeable and non-rechargeable batteries. At the moment, the lithium-ion battery is most relevant for everyday use. This is what Apple uses; its home page states that "[c]ompared with traditional battery technology, lithium-ion batteries charge faster, last longer, and have a higher power density for more battery life in a lighter package." Good reasons to run cell and mobile phones, tablets, laptops, and even Tesla cars on lithium-ion batteries.

A quick note on vocabulary: I'm only going to discuss *rechargeable* batteries, or secondary batteries, in this chapter. Nonrechargeable batteries can also be known as primary batteries. A battery provides a device with power, a flow of electrons, making it a portable source of electrons. Batteries are constructed based on the principle illustrated at right.

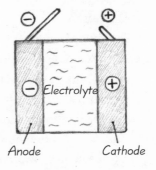

Basic principle of all batteries

The most important components are the *electrodes*, one at the positive pole and another at the negative pole. The positively charged electrode is called the *cathode*, and the negatively charged electrode is called the *anode*. Imagine that these two poles are connected by a conducting cable that runs through the phone and provides the components with power. So the electrons flow from the anode, through the phone, to the cathode. The two electrodes are also connected within the battery via an *electrolyte*. "Electrolyte" is a collective term for liquids or solids that can conduct electrical charges. However, electrolytes don't conduct electrons, only positive and negative ions. Incidentally, humans are also made up largely of electrolytes in the form of water containing all kinds of charged ions.

The basic recipe for a battery is a cathode, an anode, and an electrolyte. The chemical substances used for these three components depend on the type of battery.

In a lithium-ion battery, the cathode is often a compound of lithium, oxygen, and another metal such as cobalt. This would make *lithium cobalt oxide*. Cobalt and oxygen atoms form layers, and the lithium ions are embedded between these layers.

The anode is made up largely of *graphite*, so carbon. Graphite also has a layered structure.

When I insert the charger cable into my phone, I add an inverse voltage to the battery from the outside. At this point, something confusing happens, at least linguistically. We've only just learned two new terms—anode and cathode—and already we need to swap them over. During the charging process, the positive electrode made from lithium cobalt oxide is called the anode, and the negative electrode made from graphite is called the cathode—the opposite way round to when the battery is draining. Why do chemists make their language so complicated? It's something to do with the chemical reactions taking place in these electrodes. We'll come back to this, but for now it's probably easier to refer to the positive and negative poles.

When I charge my phone, the negative pole is loaded with electrons that make their way into the graphite material. Charging it with electrons creates a surplus of negative charge. In principle, this makes charging difficult; identical charges repel one another (negative and negative or positive and positive). Pumping the electrode full of negatively charged electrons is therefore not easily done. Opposite charges cancel each other out—and this is where the positively charged lithium ions (which give the lithium-ion battery its name) come into play. They steal away from the

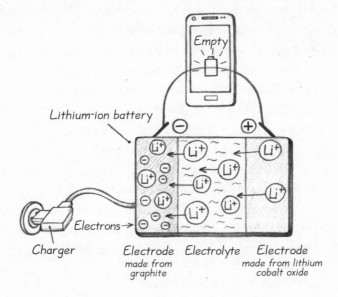

positive pole—from the lithium cobalt oxide electrode—move through the electrolyte to the graphite electrode, and take delivery of the negative electrons at the negative pole. This *charge balancing* allows the battery to be charged until it's full of electrons.

Now, the electrons don't just sit around; chemically speaking, they're welcomed with open arms. For the chemists among you, the chemical equation is $Cn + xLi+ + xe-$, which becomes LixCn. Don't worry, nonchemists really only need to focus on one part of this equation, namely the "+ xe-." The "x" simply stands for any number, and the "e-" stands for an electron. This means that electrons are incorporated into this chemical reaction. This is called *reduction. The term "reduction" is always used when electrons are gained by an element, ion, or chemical compound.*

Now that I've filled my negative pole with electrons and the positive pole has no electrons, I establish an electric voltage between the two poles. Let's imagine the electric voltage as a dam in a river. Charging the negative pole with electrons is like moving water from the bottom of the dam to the top. When I open the floodgate, the water surges down like a waterfall. This is exactly what happens as soon as I remove the charger. When I use my phone, the chemical reaction reverses and the electrons gained are released again. They surge through the phone as an electric current. This is *oxidation*, the counterpart to reduction. Therefore, *oxidation is a chemical reaction in which electrons are lost.*

While the released electrons fuel the phone, the lithium ions travel back along their familiar path through the electrolyte. The electrons and lithium ions meet again at the other side, at the positive pole. Once all the electrons have arrived at the other side, the battery is empty—and the whole process can begin again!

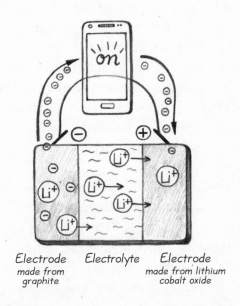

Electrode
made from
graphite

Electrolyte

Electrode
made from lithium
cobalt oxide

Because the lithium ions move back and forth between the two electrodes every time the battery is charged and drained—rocking back and forth, so to speak—the whole thing is known as the rocking chair principle.

LET'S RETURN NOW to the linguistic confusion surrounding "cathode" and "anode." In a battery that can't be charged, the cathode is always the positive pole and the anode is always the negative pole. This is only partially true for batteries that can be charged; as we've just seen, charging and draining are opposing chemical processes. This means that the cathode and anode need different definitions here: *The anode is the electrode at which oxidation happens. Accordingly, the cathode is the electrode at which reduction happens.*

When the phone is charged, reduction takes place at the negative pole and oxidation takes place at the positive pole; when the battery drains, this is reversed. Oxidation and reduction always occur in parallel and never without the other; there is always a giving (oxidation) and taking (reduction) of electrons. This is why we have a term to cover them both: *redox reactions*. Congratulations, you've completed your introduction to redox chemistry!

MANY PEOPLE SAY we should let our phone batteries drain as much as possible before charging and make sure not to charge them for too long. This is true for nickel batteries, such as those used in TV remotes, and for old lead batteries. But modern lithium-ion batteries can be left connected to chargers for as long as you like; newer batteries are built in such a way that they stop charging automatically once the battery is full. Otherwise, smartphones would be pretty dangerous. And they have been, for example in 2016, when some people reported spontaneously exploding Samsung Galaxy Note 7s. The (quite literally) energized materials inside the battery are shielded from the outside world and potential reaction partners. If they're subjected to extreme heat, the battery casing is damaged, or a construction fault leads to overcharging, then things can get unpleasant. Combined with the flammable solvents used to make the electrolyte, they form the perfect mini-bomb.

Current battery models are still not good enough for John Goodenough, co-developer of the lithium-ion battery, who is working on an even safer version, its star attraction a glassy, solid electrolyte. Don't worry, we don't need to wait in fright for this or other battery models to enter the market. The exploding Samsung model was the result of a production error that hopefully won't happen again. Imagine what would happen if the massive lithium-ion battery in a Tesla exploded...

In any case, heat isn't good, among other things, because chemical reactions can be accelerated by higher temperatures. So if your phone overheats, it can lose its charge faster. Keeping a phone or laptop battery cool can extend its total life span. And you should always carry a charger. Lithium-ion batteries last longest if you keep them more or less fully charged as often as possible. Every time the battery drains, the materials wear a little more and the battery's performance decreases. The fuller the battery before you recharge it, the longer it will last. So always keep your laptop connected to the power supply if possible and charge your cell or mobile phone as often as you can. If you're out and about and your battery is low, it's better to switch off your phone than to let it run empty.

Which is why I've just switched off my phone, although now I have to switch it straight back on again.

"I'm here," I text Christine.

"At the side," she replies.

Christine usually lets me in through the side entrance because Mr. Lässig is on my case. Mr. Lässig is responsible for safety at the institute, including its laboratories. His name might be the German word for "casual," but he is anything but. He'd be particularly displeased today because I'm wearing sandals and shorts—not because he's a creep who would hit on young women on trains, but because safety requires anyone entering a lab to wear closed shoes and full-length pants. Naturally, I'm not allowed to touch anything in the labs, not even as a trained chemist. That privilege is reserved for institute employees who have completed Mr. Lässig's safety training.

Because Christine also works weekends, I can sometimes repurpose her lab as a film set. Pouring liquid nitrogen onto the floor and moonwalking through the resulting mist looks great on camera (with closed shoes and full-length pants, of course). Thankfully, Mr. Lässig doesn't watch YouTube.

Unnoticed by Mr. Lässig, I slip through the side entrance and head for Christine's office as usual. But she grabs my sleeve and says, "No, Dino's in there."

Baby Dinosaurs and Monster Burgers

CHRISTINE STEERS ME away from her office and into the cafeteria. Like most rooms in the institute, the cafeteria is a state-of-the-art glass box. It looks sophisticated, but feels like you're constantly being watched, which is probably why it's almost empty; nobody wants to get caught slacking. But everyone knows how hard Christine works, so she can afford to take a coffee break.

Even the big laboratories invite people to look in from all angles. I'm sure the press are thrilled to see so many panes of glass—you can draw formulas on them, so they're great for photos. The staff must feel like they're under constant surveillance. Whenever I see the white coats moving around behind the glass, I can't help but think of lab rats. The head of the institute, Prof. Dr. Dr. hc. Karl Kaussen, commonly known as "King K," is very proud of all the glass. Last year, I joined a few colleagues from local and university media outlets for an exclusive tour of the "innovation spaces."

"There's so much daylight; isn't it great? You know, daylight stimulates the release of serotonin, the happiness hormone. That's why our employees are always in such a good mood!" King K said as he guided us through laboratories flooded with light. The hardworking PhD students nodded and smiled like good little boys and girls—and that was when they first reminded me of lab rats. King K is a charismatic man, but Christine had already told me that behind the media-friendly facade, he was a real tyrant, which was why I recognized the unbridled fear behind the smiles. King K's mere presence is enough to provoke a fight-or-flight response in doctoral students (mainly flight...).

His comment about *serotonin* being a "happiness hormone" was a bit too simplistic. Serotonin is a hormone and, like most hormones, the molecule can affect the body in many different ways, which in turn are linked to a whole range of complex chemical reaction chains. I'm going to emphasize this again: I don't want you to read this book and get the impression that a hormone can be reduced to one single effect, like melatonin being a "sleep hormone" or cortisol a "stress hormone." Among its many functions, serotonin influences our mood, which is why for decades it has been connected with depression.

It stands to reason that we would search neurochemistry for the causes of mental disorders. One of the philosophies of psychology is that *everything psychological is simultaneously biological*. (I'm going to be cheeky and add that everything biological is simultaneously chemical. It's just that we understand only a tiny fraction of the chemistry running our body.) The neurons—the nerve cells in our brains—communicate by sending molecules back and forth. Nerve cells are linked via contact points called *synapses*. They don't actually make physical contact—there's a tiny gap between them, known as the synaptic cleft. Neurotransmitter molecules are fired across this gap from one nerve cell to another and park in the *receptors* on the other side. You can actually visualize receptors as parking spaces reserved for certain molecules. When a molecule docks at a receptor, the signal is either activated or inhibited.

So *neurotransmitters* fulfill the same role as *hormones*; they are messenger substances. Whether a molecule is called a neurotransmitter or a hormone depends on where it's released in the body. If it's released at the synapses, then it's a neurotransmitter. If it's produced in a gland, such as the pineal gland or adrenal glands, then it's a hormone. Serotonin can be both a hormone and a neurotransmitter.

In the 1970s, people began to suspect that depression could be caused by low serotonin levels. There was evidence that increasing the serotonin level in the brain helped combat depression. So, for a long time, there was a fairly simple explanation: depression is triggered by a chemical imbalance in the brain, and we can combat depression with medications that increase serotonin levels, for example. However, it would be far too simplistic to reduce a mental disorder such as depression to the absence of one little molecule. Just because serotonin helps against depression doesn't mean that depression is caused by a lack of serotonin; just because aspirin helps get rid of a headache doesn't mean that the headache was caused by a lack of aspirin. And yet, in practice, antidepressants that boost serotonin levels do help many people who suffer with depression. Is there a placebo effect here? Are we fighting only the symptoms, or the causes too? The complex link between serotonin and depression remains controversial. That's how science goes. Most researchers will probably become disillusioned after a few years. Science and technology have brought us a long way, but searching for new findings is the work of the devil. Results contradict themselves or can't be reproduced. Science isn't always clear and logical.

The "blind men and the elephant" is a good allegory for science and research. A group of blind men examine an elephant, a creature they have never come across before. Touch is the only way for them to visualize it, but each man touches only one part of the elephant. One feels its pointed tusks, another its long trunk, another its large ears, and so on. When they discuss it, they realize they each have a totally different mental picture. One is adamant that the elephant is bony and spiky; another disagrees completely. Scientific research is often like this, particularly when the elephant is a topic as complex as depression. To have a chance of figuring out what's going on, you must first be able to understand the elephant as a whole by assembling all observations, even those that initially seem contradictory, to form a meaningful picture. Even then, we have only explored the elephant through touch. We remain blind.

When I discuss these issues with friends, they ask whether I'm trying to interest them in science or scare them away. Well, sticking with the allegory, without science we wouldn't just be blind, we wouldn't even be able to touch.

MOVING ON, LET'S turn our attention to brain chemistry, a well-researched area. Christine and I are sitting in the cafeteria, supplying our brains with *caffeine*. Our brains have receptors not only for neurotransmitters, but also for caffeine, for example. This isn't deliberate, though; caffeine looks confusingly similar to another of the body's molecules, *adenosine*.

Caffeine

Adenosine

The chemical structures may not look all that similar right now, but they do to the adenosine receptors in the brain. It isn't actually about how they look, but how well a molecule fits into a receptor. Adenosine fits perfectly into the adenosine parking space—and caffeine just happens to fit in too.

Normally, adenosine lets us know when we're tired. The more adenosine molecules dock in adenosine receptors, the more tired we feel. Where does the adenosine come from? Appropriately enough, it's linked to energy consumption. The more energy we use, the more adenosine is produced. Whenever our body uses energy, whether during exercise or simply through thinking or breathing, it requires a molecule called *adenosine triphosphate* (more for exercise, less for breathing).

Adenosine triphosphate

Adenosine triphosphate, also known as ATP, is our body's unit of energy. I think the long name will be more informative here—it shows that when adenosine triphosphate loses its three phosphates, only adenosine remains. Remember it like this: the more ATP molecules we use, the more adenosine is produced (a biologist would say differently, but we don't need to make things unnecessarily complicated). And the more adenosine blocks our receptors, the more tired we become.

Unless, of course, we resort to coffee! When we consume caffeine, it takes around fifteen minutes for the caffeine molecules to arrive at and dock in the adenosine receptors. They can even force out and replace any adenosine molecules already in there. The parking space is now blocked with caffeine, but the receptor doesn't notice. Because it doesn't "see" any adenosine, it thinks it's free—and we think we're awake!

DESPITE THE COFFEE, Christine is a bit subdued—she's just found out that her paper has been rejected. *MythBusters'* Adam Savage was right when he said, "The only difference between screwing around and science is writing it down." An experiment isn't science unless it's been properly documented and analyzed. Once your investigations lead to new findings, you can publish the whole thing as a "paper."

This one little word harbors a long and often frustrating process that is almost a science in itself. Once you've collected your results, gotten them in order, and written them down, you can submit your paper to a scientific journal. The journal's editor then selects reviewers, usually professors at other universities who work in the same field; the author doesn't know who these reviewers will be. The reviewers read and evaluate the manuscript and decide whether it's worthy of inclusion in the journal. Before publication, you'll need to revise the paper based on their comments, often adding more data or experiments. This process is known as *peer review,* and aims to ensure high-quality scientific publications.

A paper may be rejected by several journals before being published (after careful revision, of course). The first reviewer might think it's great, but the second reviewer might not be convinced at all. This is what's just happened to Christine. She shows me the second reviewer's comments.

"Read that. The guy obviously hasn't understood the section at all."

"It could be a woman!" I chide.

"Whatever, they haven't understood it."

"But that's good," I say. "You might just need to rework the text."

Experts can easily forget how much knowledge is required to understand their work—even when writing a scientific paper, not just when explaining chemistry to someone at a party. Scientists tend to use excessively complicated language without realizing, even among themselves. The issue is compounded because many people don't admit to being confused. It would be embarrassing if you turned out to be the only idiot who didn't get it. Academics can't allow themselves to be humiliated like that. Attending lectures at the start of my doctorate, I was convinced that I must have been incredibly stupid; I hardly understood anything. I mean, the others didn't seem to have any questions. I soon realized that most of us were in the same boat. Today, I know that it's usually just the lecture that's incomprehensible.

When you give an incomprehensible lecture, you waste the audience's time; when you write an incomprehensible paper, you shoot yourself in the foot. It's always best to run it past someone else to check whether everything makes sense; eventually, you can't see the wood for the trees.

"Can you take another look at the paper?" Christine asks me.

"Sure," I say. Making science understandable is my specialty, after all.

NORMALLY, WE'D SIT in Christine's office. Unlike most of the "innovation spaces," her office is only half glass. It's always given us a little privacy when I've visited in the past, but recently Christine started sharing her office with Dino. His real name is Torben; he's one of King K's postdocs. He looks a bit like an oviraptor, which is why we secretly—but affectionately—call him Dino. He's strangely bony and when he moves he hunches forward slightly. I didn't think the oviraptor comparison worked at first; oviraptors were wily and impudent, stealing eggs from other dinosaurs. Torben would never do anything like that. Then Christine explained that the name "oviraptor," which translates to "egg thief," is based on a misconception. An oviraptor skeleton was found over a nest, leading to the false conclusion that it was pilfering the eggs. They later turned out to be its own eggs. Should have been obvious really. So oviraptors might have been really lovely dinosaurs. Like Torben.

Torben is a particularly extreme kind of nerd, and I don't say that lightly. You'll have noticed by now that I'm a pretty serious nerd myself. I struggle not to view the world through scientific eyes. When I absolutely have to, I can put myself in the unfortunate shoes of nonscientists and blend in with them completely. Torben is incredibly shy, bordering on having an anxiety disorder, and I'm not exaggerating or being ironic. Social anxiety is a recognized anxiety disorder according to DSM-5 (the latest edition of the Diagnostic and Statistical Manual of Mental Disorders). I don't want to pretend that I'm capable of making psychological diagnoses, let alone remote diagnoses. I've only met Dino once, and most of what I know about him has come from Christine. Last time I was here, he'd just joined the institute. We were sitting in Christine's office at the desk opposite Dino, who looked like he'd rather be sitting elsewhere. I'd brought a pack of apple rings (dried rings of real apple, not the sour candies) and offered them to Dino. At first he didn't seem to respond at all, but a trained nerd could see that he had registered my offer and was considering it intensely. In awkward situations like this, the main thing is not to let the awkwardness show. Don't start laughing nervously if there's a long pause in the conversation; simply relax and wait to see what the other person does (remember, be a noble gas!). If you think about it, many conventions for social interaction serve no logical purpose. Yet many nerds can only think in logical terms. So they remain stuck in a bubble, awkwardly negotiating the world of small talk and other social conventions. Many people will label them freaks after just a few interactions, and that label sticks. If you can be open, it's often just a matter of time before the bubble bursts. Christine and I have seen it happen quite a few times. And sometimes, even the quietest dinos turn out to be the coolest people.

So I held out the pack of apple rings until Torben took one. Christine (who had already started to take him under her wing) jokingly asked, "Or are they too healthy for you?"

Still dutifully holding the apple ring, Dino replied calmly, "I'm highly allergic to apples."

"How's Dino?" I ask.

"I'm a mom now," Christine sighs. "You know how when a baby duck hatches, the first moving object it sees becomes its mom and it follows that object wherever it goes? Dino is my baby duck, and I'm his mom."

"Baby dino," I say, wondering whether baby dinosaurs did the same as ducks. The closest living relative of the T. rex is the chicken.

"It's 2:30, the cafeteria is about to close, and he still hasn't eaten because I haven't eaten."

"Wow."

"Yep. Brace yourself, because from now on it looks like I'll be taking him with me wherever I go."

I picture Christine and Dino fifty years from now, knitting by the fire in rocking chairs.

We quickly make our way to the cafeteria before it closes. Christine wants to test what happens when she doesn't take Torben with her. Our route takes us past Christine's office. We peer through the partially glazed wall. Dino is sitting at his computer and doesn't seem to notice us, but Christine is convinced he'll follow us soon. Even though I'm here too? I mean, I almost poisoned him last time.

WHILE WE'RE ON the subject of poison, last year a group of around thirty students stood at the main cafeteria entrance, brandishing posters and protesting against preservatives. I don't think a cafeteria without preservatives would be a very good idea. I'm a big fan of fresh food and cook with fresh ingredients whenever I can, but I'd be glad to see the occasional preservative in a cafeteria that serves thousands of people every day. After all, the world is full of bacteria, fungi, and other microorganisms, and they need something to eat too—our food, for example. I wouldn't be opposed to sharing my food with the little critters (they wouldn't eat much) if it weren't for the fact that they'd turn it into a disgusting, smelly mess, let alone the higher risk of infection and poisoning. Everyone's heard of salmonella, but the list goes on. Botulism sounds like a branch of philosophy, but it's actually a life-threatening form of food poisoning.

In addition to these bacteria, we have the purely chemical processes that can spoil our food (obviously, the toxins in bacteria are produced by chemical reactions in the bacteria's metabolism, but you know what I mean). The classic chemical reaction that spoils food is *oxidation*, which we discovered when talking about phone batteries.

Oxidation can be defined in various ways. In a phone battery, oxidation is a chemical reaction involving the loss of electrons. This is the general definition. You can also define oxidation more precisely and literally, namely as a chemical reaction involving oxygen. When fats react with oxygen—when they oxidize—they become rancid and inedible. When you slice an apple, it's the oxidation of polyphenols that turns the apple brown. You might have noticed that different varieties of apple turn brown at different rates. Newer varieties like the Granny Smith and Golden Delicious are generally lower in polyphenols. They look nicer, but are bad for people with apple allergies, like Dino. The higher the polyphenol level in an apple, the more an allergic person can tolerate it.

Oxidation usually requires *enzymes*. Enzymes are part of the protein family and can be found everywhere—in humans, animals, plants, fruits. Their chemical structures are extremely diverse and they work in highly varied ways, but one thing they all have in common is that they serve as *catalysts* for chemical reactions. This means that enzymes can help molecules to do what they want, but struggle, to do—like a nice young man who helps an elderly person get off a streetcar. Other enzymes are like a good dating agency, bringing the right reaction partners together. Others are like kitchen helpers who chop the ingredients to speed up the cooking.

Enzymes are an impressively diverse class of substances. Apples contain an enzyme called polyphenol oxidase, or PPO, that helps them turn brown. The name clearly indicates that PPO is specifically responsible for the oxidation of polyphenols, and its ending ("-ase") is typical for enzymes. Our metabolism wouldn't work without enzymes either; most chemical reactions that keep our bodies alive wouldn't happen at all without enzymes or would be far too slow. This is why I can't tolerate alcohol; one of the enzymes responsible for breaking down alcohol doesn't work in my body. I'll come back to this in Chapter 13.

So rotting food is nothing more than a series of unwanted chemical reactions. Fridges and freezers offer effective physical preservation—the lower the temperature, the slower most chemical reactions occur. There are also various chemical strategies to protect food from spoiling; you can either make life difficult for the microbes or the enzymes or get rid of the oxygen. There's more than one way to skin a cat.

Let's start with the *oxygen*. It's not easy to protect against oxygen on this planet, given that it's part of the air (not that we're complaining about that). Manufacturers can either vacuum-pack their food products or put the packaging in a protective gas environment (like the argon or nitrogen in Chapter 6). The protective gas used for food products is usually a low-oxygen blend of nitrogen and carbon dioxide (CO_2). You can't remove all the oxygen from food packaging, but the less contact with oxygen, the less oxidation. Putting chocolate spread on a slice of apple also creates a pretty solid protective layer against oxygen (a little unhealthy tip for you there). You've probably heard of the healthier option of squeezing lemon juice over the apple; lemon juice contains vitamin C, an antioxidant.

Vitamin C
(ascorbic acid)

Everybody talks about *antioxidants*; the word even gets bandied about in cosmetics commercials. The chemical—and literal—definition of an antioxidant is a substance that prevents oxidation because it likes to react with oxygen and does this well. A true martyr, an antioxidant throws itself into the firing line and cries, "Leave the polyphenols alone! Take me instead!"

LEMON JUICE IS also acidic. And *acid* obstructs many enzymes. Enzymes are huge, complex molecules, ingeniously shaped and folded like some sort of funky origami. Their three-dimensional structure is the secret to their precision. They can bring two specific reaction partners together by, for example, putting one in a perfectly pre-folded pocket and holding it there to provide the other with better access. But acid can *unfold* enzymes, meaning that they lose their three-dimensional structure and their effect as a chemical catalyst. This is why pickled gherkins last so long. The acetic acid preserves them. There are even bacteria that, with acid, help to preserve foods. I need to emphasize here that not all bacteria are bad, and sometimes there's a fine line between good and bad. Milk goes sour due to lactic acid, which is produced by bacteria. Lactic acid is specifically used to make yogurt; certain lactic acid bacteria are added to the milk, turning it into acidic yet tasty products that last longer. The same trick is used to make sauerkraut, which is famously long-lasting.

Acetic and lactic acid bacteria aren't the problem—the problem is the CHEMICAL preservatives!

First, let's agree to call them "artificial preservatives"—preservatives created in a lab. Look at the ingredients listed on a pack of food and you'll find artificial preservatives indicated by mysterious "E numbers." The most important are acids such as *sorbic acid* and its salts, *sorbates*. You'll find them listed under E 200, E 202, and E 203. Their chemical structure is a little like that of fatty acids and their salts, which we use as soaps. They are metabolized like natural fatty acids in foods, and you don't need to worry about toxicity. Sorbates and sorbic acid also have a fairly unobtrusive taste and smell.

Sometimes other acids are needed too, such as *benzoic acid* and its salts, *benzoates*. They can be found under numbers E 210 through E 213. They target yeasts and molds that run rampant even at low pH levels (i.e., in the presence of other acids) and are used in acidic foods like mayonnaise, canned fish, pickles, and soft drinks. Benzoic acid isn't as free from scandal as sorbic acid. There's a suspicion that benzoic acid may make children hyperactive, but as the European Food Safety Authority has established, there is no scientific evidence to back this up. Benzoic acid and benzoates are officially classed as safe.

As a chemist, I see some objections that particularly annoy me. Again and again, I read that benzoic acid is a naturally occurring substance and therefore harmless, but benzoates are artificial and therefore dangerous. This is nothing more than mindless chemism—benzoic acid and benzoates are two different forms of the same molecule! Like fatty acids, benzoic acid is a carboxylic acid. Carboxylic acids always have two different forms, acid and salt. Just like fats and soap, the acidic form is difficult to dissolve in water, while the salt form (ionic, charged) dissolves well. Which of the two forms occurs simply depends on the pH value of the environment.

In an alkaline environment (high pH), we have more benzoate; in an acidic environment (low pH), we have more benzoic acid. If you add benzoate to an acidic food—and this is exactly how it's used—then part of the benzoate is inevitably converted into benzoic acid. So the whole "natural" versus "artificial" thing is nonsense!

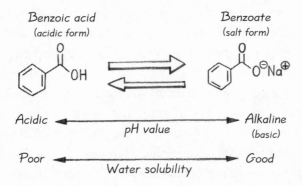

If ingredients are listed by quantity, then the preservative E numbers usually come last. We don't need much of them, but toxicity must be examined as thoroughly as possible. The technical capabilities of laboratories are constantly evolving, and it's now much easier to collect data for long-term studies. It's important to continue researching additives that have already been approved so that they can be reviewed regularly based on new, solid data. You already know that most research is complex and can take a long time. If there's any suspicion of danger to health or safety, incomplete results should be taken seriously. This allows alternatives to be considered at an early stage. But panic(-mongering) and chemism (or chemism-related panic-mongering) are not the solution, particularly if no useful alternatives have yet been found.

Chemism or not, I can totally understand why people refuse "artificial" ingredients in favor of "natural" ingredients. I'd rather have fresh food too; it tastes much better. But just because the label says something is "natural" doesn't automatically mean it's fresh. This becomes clear when we compare natural and artificial flavoring agents. If you know which molecules create the flavor of a natural fruit, you can either extract them (from a natural source) or produce the molecules yourself in a laboratory (CHEMISTRY!). Provided they have the same chemical structure, there's no difference between molecules from nature and molecules from a lab; it's just that nature is a far more accomplished chemist than all human chemists put together. Flavor often traces back to a sophisticated blend of molecules, while artificial flavors often have a simpler composition. This also means that artificial flavors are just as safe as natural ones, if not safer, because every single ingredient in the artificial flavor has been identified and tested.

Incidentally, just because a label says "natural flavors only" doesn't mean that these flavors come from the fruit in question. For example, "natural coconut flavoring" is often based on a molecule called massoia lactone from the bark of the Massoia tree. It's natural, but it's not coconut.

If you value food sustainability, then you mustn't demonize all artificial ingredients. Citric acid, which is also used as a preservative, occurs naturally in lemons and other citrus fruits, but there wouldn't be enough citrus fruits on the planet to meet the demand for citric acid. So I commend artificially manufactured citric acid; it really doesn't matter whether the molecule is produced by a plant or a human chemist.

See? Chemistry isn't always so wrong, is it?

ON THE SUBJECT of chemistry, do you remember the mysterious case of the McDonald's cheeseburger that never went bad? In 1996, Karen Hanrahan bought a McDonald's cheeseburger, and she may well be carrying it around with her today. Several years (years!) after she bought the cheeseburger, it was still intact and showed no signs of rotting—it basically still looked like a McDonald's cheeseburger. Maybe it still does? The last I heard of the immortal burger was back in 2012. Naturally, the story made headlines for years. What the hell was in this monster burger? What cocktail of chemicals creates something like that?? What are we even eating???

The answer is both a relief and a slight disappointment. It didn't contain any monster substances; McDonald's burgers are simply very dry. In just a short time, the burger dried out so much that there wasn't enough water left for the bacteria and fungi to do their thing. Karen Hanrahan was truly committed to warning the public against fast food, and the dried-out cheeseburger was her most impressive prop. She used it as "evidence" of the terrible chemistry of preservatives. I support her intentions in principle—we should all keep our fast-food consumption to a minimum—but it makes me angry when scientific findings are twisted to promote them. There are so many other wonderful, scientifically proven reasons to step away from junk food!

What this cheeseburger does show, however, is that oxygen isn't the only thing that spoils our food; water does too.

You can preserve food by simply drying it out completely or adding loads of sugar or salt (also not exactly healthy). Sugar and salt both soak up water; they're so hydrophilic that they're irresistible to water molecules. The water molecules cozy up to the salt ions or sugar molecules, meaning they are no longer freely available for the microbes.

THE LITTLE STUDENT protest at the cafeteria didn't stop preservatives from being used, but they are shown more clearly on the menu now. My fear is that simply listing the preservatives and mysterious E numbers will worry people even more. Most people know far too little about the preservatives that form part of modern life—which is why I always say that we need to understand chemistry better, all its nuances, risks, and possibilities.

Obviously, we should eat fresh, unprocessed, and local ingredients whenever we can. We can also be pragmatic and remember that centuries of research have given us the luxury of buying food in a supermarket and the "luxury" (if you can call it that) of eating in a cafeteria, and that this would be inconceivable without preservatives.

Preservatives are just one of many examples. Generally speaking, we have so many reasons to thank chemistry and artificially manufactured substances. We're scared of poisons and unnatural chemicals, but unaware of chemistry's countless positive achievements that make everyday life easier and even save lives—be they medications, power cable insulation, or volumizing hair spray.

Sometimes, the fear of chemistry reminds me of the fear of vaccinations. Vaccines work so well that we've forgotten all the horrible diseases they've wiped out. We take it for granted that we live without smallpox, diphtheria, and polio. We don't appreciate it. Instead, we worry about potential, rare side effects that are nowhere near as common or dangerous as the diseases we're fighting.

I admit that comparing vaccines and chemistry is a bit iffy. You can't say that chemistry is fundamentally less dangerous than it is useful. I just want to warn against generalizing; it's such an easy trap to fall into. I do the same thing, although in the other direction. One time, I went to the pharmacy to get something for a cold. The pharmacist offered me two products, one herbal and one chemical. I instinctively chose the chemical option because I wanted something that would actually work—then realized that I'd automatically assumed the chemical product would have a stronger effect, without asking for more information. I had to give myself a slap on the wrist; an artificial substance from a laboratory isn't automatically more effective than something from a plant.

If we are all aware of our prejudices and take a long, hard look at ourselves every now and again, we can be fairer to both chemistry and nature and make better decisions.

CHRISTINE PRODS ME and nods toward the cafeteria checkout. There he is, the oviraptor, trotting over with his tray.

"He will follow yooouuu," I sing to the melody of "I Will Follow Him."

"I should just ask if he wants to move in with me." Christine's words are sarcastic, but I know she's taken Dino into her heart. "I'm actually pretty proud of him for not letting you intimidate him."

"Well, I'm a nice person," I say.

"Well, I'm a good mom," Christine says.

Torben sits at our table and something wonderful happens. He looks at me, grins—and puts an apple on my tray. What a comeback. In my mind, I see everyone in the cafeteria rising to their feet to give him a standing ovation. There's a ticker-tape parade, banners and flags flying, crowds of people carrying Dino on their shoulders. Christine and I beam at Torben and he starts to laugh.

"The bubble has burst," Christine whispers.

Covalently Compatible

"KING K COLLARED me before I could leave the office," Dino says in his wonderfully serene manner.

"What?!" Christine yells. Her maternal instinct has clearly been roused. I'm starting to think *she's* the one who started this whole mother/baby dinosaur thing. "What did he want this time?"

"He says we have to remove the plants from the office. Apparently they don't fit with our corporate identity."

"With WHAT?" Christine and I shriek in unison. Sometimes, I'm glad to be out of academia; it messes with your head.

"And what did you say?" I ask Torben.

"I asked him if I should get rid of the plants then and there," Dino says calmly. "He got pretty angry and said, 'You have a doctorate; don't ask such stupid questions.'"

It's clearly been a while since anyone's said "no" to King K. And nobody's going to tell him that his office plant idea is a load of garbage. That's how it goes when you're the top dog. There are hierarchies everywhere, of course, but they're particularly insidious in academia. In a company, every boss has a boss. The executive board members have shareholders or an employee organization breathing down their necks. In theory, the only person a professor answers to is God, and since scientists tend to be atheists, they have absolute power.

Obviously, there are lots of great professors. The PhD advisor I shared with Christine was wonderful. If he hadn't been so wonderful, Christine would probably be working in industry right now and earning a lot of money, which would have been a great loss to science. Sadly, in a world of fixed-term contracts, mediocre salaries, and ridiculous pressure to succeed, not enough people manage to provide even a modicum of emotional support and appreciation. The following episode is a particularly striking example.

In 1996, the same year in which Karen Hanrahan bought her immortal cheeseburger, chemist Erick M. Carreira wrote a letter. At the time, Carreira was an ambitious chemistry professor at Caltech, the California Institute of Technology, one of the world's most prestigious universities. This private letter was addressed to Guido, one of his postdocs, but was somehow leaked to the public and quickly made the rounds:

Guido:

I would like to provide for you in written form what is expected from you as a member of the research group. In addition to the usual work-day schedule, I expect all of the members of the group to work evenings and weekends. You will find that this is the norm here at Caltech. On occasion, I understand that personal matters will make demands on your time which will require you to be away from your responsibilities to the laboratory. However, it is not acceptable to me when it becomes a habit.

I have noticed that you have failed to come in to lab on several weekends, and more recently have failed to show up in the evenings. Moreover, in addition to such time off, you recently requested some vacation. I have no problem with vacation time that is well earned, but I do have a problem with continuous vacation and time off that interferes with the project. I find this very annoying and disruptive to your science.

I expect you to correct your work-ethic immediately.

I receive at least one post-doctoral application each day from the US and around the world. If you are unable to meet the expected work-schedule, I am sure that I can find someone else as an appropriate replacement for this important project.

Sincerely,
Erick M. Carreira

The letter's publication caused outrage—understandably—but the author was only being honest. Many chemists were more shocked by its bluntness than by the expectations expressed, which are not uncommon in academia. I've heard a whole range of stories, but that's a subject for another book.

When Carreira wrote the letter, he was in a similar position to Christine now: thirty-three and an associate professor in the "make or break" phase of his academic career. Today, he's one of the world's most renowned organic chemists. I've heard he's become pretty cool and laid-back.

Christine and many of the other young scientists I know give me hope for the future. I just hope that a few years of university work doesn't turn them into ogres too. Christine regularly gives me permission—actually, she orders me—to slap her if I ever see signs of her treating her students cruelly.

INTERPERSONAL RELATIONSHIPS ARE the most important thing of all. It's just like in chemistry; atoms and elements are all fascinating in their own way, but things only get truly interesting when atoms bond to form molecules and enter into chemical reactions with one another. Some chemical bonds are similar to human relationships. Our circle of friends includes an inseparable couple known as the "Steffis" (their names are Steffi and Stefan—yes, really). Although that sounds nauseatingly harmonious, they're actually the exact opposite. Some might say that they constantly fight, but strictly speaking it's only Steffi who fights; Stefan endures it all in silence. Their relationship seems very unbalanced; Steffi takes a lot, Stefan gives a lot. Christine thinks Stefan's getting a raw deal, but I suspect he likes things just as they are. They've been together for a long time and (I believe) are happy in their own way. Christine would describe the Steffis' relationship as a "metastable bond," but I see it as more of an "ionic bond."

There are three different types of chemical bond, or at least three types worth learning: ionic bonds, atomic bonds (also known as covalent bonds), and metallic bonds. Two atoms are always bound by the same thing: electrons. *Chemical bonds are formed through the sharing of electrons.* To be more precise, the valence electrons are shared among the atoms in a chemical bond (this should make sense after our discussion in Chapter 2). Why atoms bond at all is due to the octet rule, for example. But the kind of bond that's formed depends on how exactly the electrons are shared between the binding partners.

In an *ionic bond*, one binding partner gives an electron to the other. They may give several electrons, depending on what the octet rule requires, for example. We've already seen this in the sodium fluoride in toothpaste and in our cooking salt, sodium chloride. Positive and negative charges are created—cations and anions (mentioned in Chapter 2), which attract one another. The same often happens in relationships, bonds that we enter into as people. It's always said that opposites attract—voilà, ionic bonds!

Unlike the Steffis, however, we can't visualize an ionic bond like sodium chloride as individual twosomes of sodium and chloride. The electrostatic attraction between the positive sodium and negative chloride works radially, going in all directions from each ion. Each sodium ion surrounds itself with chloride ions, and vice versa. This results in an orderly, three-dimensional structure known as the *ionic lattice*. The ionic lattice for cooking salt looks like this:

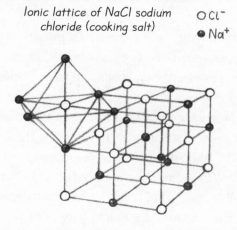

Ionic lattice of NaCl sodium chloride (cooking salt)

○ Cl^-
● Na^+

In Chapter 2, I described this kind of bond as a "model marriage," but that analogy might not quite fit anymore. An ionic bond is an unbalanced relationship; one gives and the other takes. But they're happy. Like the Steffis. One wants nothing more than to give, and the other nothing more than to take (just like our redox reactions in Chapter 6). So maybe this is a perfect marriage after all, although not necessarily a good example; an outsider looking in would probably judge Steffi and pity Stefan. This principle of chemical bonds can also be applied to other interpersonal relationships—Christine and Torben's mom/Dino relationship seems pretty much like an ionic bond to me.

NOW LET'S LOOK at another type of bond, one that fits me and Christine. In Chapter 2, we looked at the bond between carbon and fluorine in Teflon. Organic bonds—bonds containing carbon—are predestined for *atomic bonds*, also known as *covalent bonds*. Personally, I prefer to call them covalent bonds; "atomic bond" seems pretty meaningless. Aren't all bonds ones between atoms?

Covalent binding partners share electrons, rather than one partner giving and the other taking. They are held together by shared electrons, not by the electrostatic attraction between charges.

You've seen a few covalent bonds in this book so far:

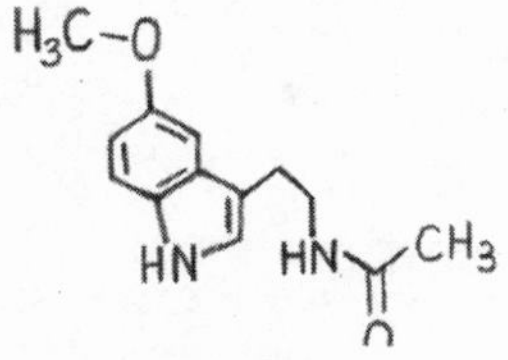

Melatonin

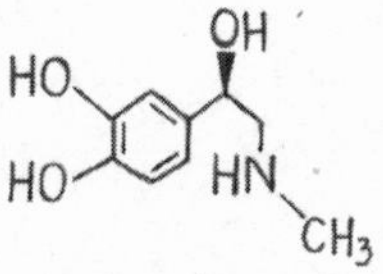

Adrenaline

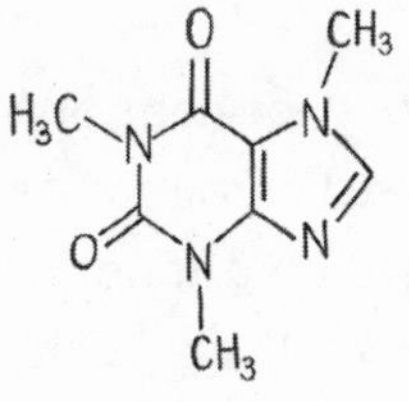

Caffeine

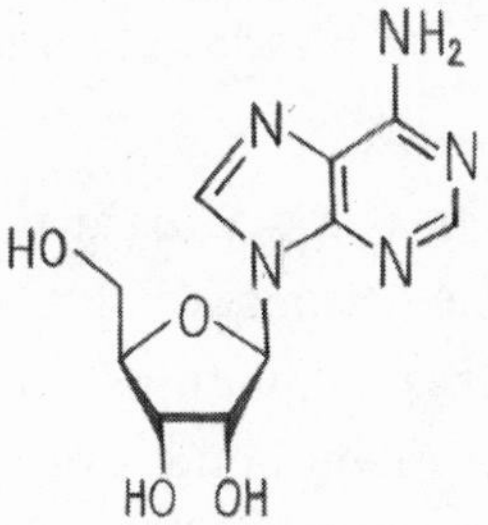

Adenosine

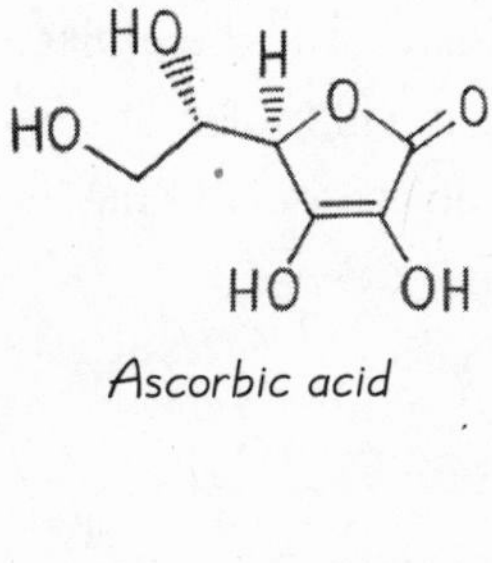

Ascorbic acid

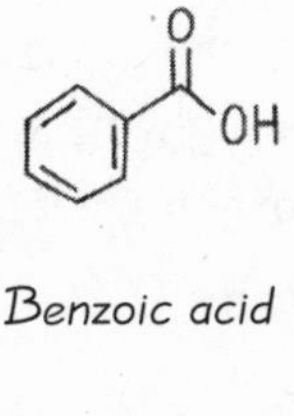

Benzoic acid

Every line in this diagram represents a covalent bond, and unless labeled otherwise, every angle represents a carbon atom. Because our lives are so dominated by carbon, most molecules contain so many carbon atoms that there's no point writing "C" everywhere (or CH or CH_2):

Simplified diagram of chemical structures using melatonin

Carbon is a master at forming covalent bonds; this is one of the reasons why life is based on carbon and why we assume that all life must be based on carbon. Unlike ionic bonds, covalent bonds don't work radially in all directions; there is one specific direction and one specific angle between two bonds. This is why covalent bonds can be used to build far more elaborate structures than ionic bonds—including our DNA; huge, complex proteins; and all the molecules you'll find in this book. Even simple, small molecules like our gas molecules only work with covalent bonds, while ionic bonds instantly form huge lattice structures.

The division between ionic and covalent bonds is, how-ever, fluid. The binding partners in a covalent bond aren't always totally equal; even when sharing electrons, one part-ner can claim more than the other. How fairly the electrons are distributed depends on the difference in electronegativity between the binding partners. *Electronegativity* is a chemical element's ability to attract the electrons within a bond. In a bond between two equal atoms, traditionally two carbon atoms, the bond is actually very fair and balanced. But in a water molecule (H_2O), the oxygen (O) atom is more electro-negative than the hydrogen (H) atoms, and therefore attracts the shared bonding electrons more strongly. This doesn't immediately make the water ionic, but the *electron density* is greater at the oxygen atom than it is at the hydrogen atoms. This creates *polarity*—similar to the positive and negative poles in a battery, the charge in a water molecule is also sep-arated, albeit to a lesser degree. Another way of saying this is that the oxygen atom has a *partial negative charge* and the hydrogen atoms a *partial positive charge*. A covalent bond with an ionic touch, you might say. This is also known as a *polar atomic bond* or *polar covalent bond*.

If the electronegativities are extremely different, one of the binding partners will attract all the electrons, and then we get an ionic bond.

I ENJOY DIVIDING interpersonal relationships into ionic and covalent bonds. Some people seek out opposites in their friends and partners, but I'm more the covalent type. I think my marriage is very covalent, and I choose friends who are as "covalently compatible" as possible. Others might find that too boring, but it has its advantages. When Christine organized my bachelorette party, some of my friends had never met before. Christine was seriously impressed at how well everyone got along, how easy it all was. It must be due to my preference for covalent friendships.

THEN THERE ARE *metallic bonds*, a whole other story. They hold gold bars and iron nails together—and, of course, our spoon from Chapter 1, in which we visualized the metallic bond as a jungle gym. You must be ready for more details by now, right?

The chemical elements can be divided into metals and nonmetals; around four-fifths of the elements on the periodic table are metals. They share an interesting form of bond that can be described using the *electron gas model*. The valence electrons aren't bound to any particular atom; instead, they move relatively freely within the metal in a similar way to the molecules within a gas. This wild community of electrons is therefore also known as *electron gas*.

So, in a similar way to covalent bonds, metallic bonds are based on shared electrons, although the metal is like a commune—everyone shares everything.

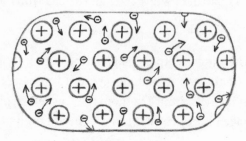

Electron gas model

When the metal atoms allow their valence electrons free rein, they themselves remain behind with a positive charge. As positive *atomic cores*, they form the framework of the metal, the *metal lattice*. The positive atomic cores and the electrons in the electron gas attract each other, similar to the cations and anions in an ionic lattice. Because the electrons can move, a metal lattice isn't as rigid as an ionic lattice. It is this feature that determines the properties of a metal.

Metals have three characteristics that all result from this form of electron gas bond.

One: Metals conduct electric current.

Electric current is nothing more than flowing electrons. Naturally, electrons can flow really well as electron gas. If you connect a metal wire to a battery, electrons are simply shoved in at one side and pulled out at the other.

Two: Metals are good heat conductors.

Now we come back to Chapter 1 and the question of why a metal spoon feels colder than a wooden tabletop. If heat conduction is nothing more than the transfer of kinetic energy between particles, then it works particularly well if the particles can move as freely as possible and collide as frequently as possible. With every collision within the electron gas, the energy is transferred from one electron to the next. Wood, meanwhile, is made up of covalent bonds— much more rigid from a molecular perspective and worse at conducting heat.

Three: Metals are easy to deform.

Now, I don't mean that metals are necessarily soft. Hardness and deformability are two different things. A metal wire is both hard and malleable. If we take a piece of wood or glass and try to bend it, it will eventually break—both materials are brittle. But a metal stick can be deformed because the atomic cores that form its basic framework aren't fixed rigidly in place. They glide past each other smoothly, cushioned by the electron gas, allowing metals to be hammered and forged without shattering.

It never fails to fascinate me how physical and biological properties can be ascribed to chemical structures. It's wonderful! And it's even cooler when you can use this knowledge to create your own molecules and materials with the properties you desire. How can anyone not think that's wonderful? How can anyone not find chemistry wonderful?? I'll never understand.

"OH NOOO," CHRISTINE says, looking mournfully at her phone.

"What?" I ask.

"Young Tesla's turned me down. He's going to King K." Christine is dejected.

"Young Tesla" is a young man who applied to do his doctorate with Christine and apparently looks like a young Nikola Tesla. He seemed to be a very good student and made a promising and lasting impression at his interview. Christine's research group is small, but she has a knack for choosing the right people—good scientists who are also team players, a combination she really values. She wants her group to have a constructive, inspirational atmosphere; she once turned down an applicant who would have fit in professionally, but not on a personal level. Young Tesla seemed perfect for her team, but Christine has to compete with eight other research groups at the institute. And most applicants want to join King K; he's head of the institute and the biggest name. Obviously a "junior group" would be risky; you never quite know how successful new groups will be. Junior professors like Christine are under great pressure themselves, so they're usually better supervisors, but there's a greater risk to students of being treated like a slave—remember Carreira's postdoc? Anyhow, I'm sure Young Tesla will soon regret his decision.

"I'm going to spend the next few years watching King K run him into the ground," Christine says.

I put my arm around her. "It's been a crappy day, hasn't it?"

"Hmm."

Dino looks depressed too.

"Come round tonight. I'll cook," I say. Christine hesitates. I know what she's feeling right now—a desire to work even harder, to make even faster progress, so that she doesn't lose more good people to King K. Who's got time for dinner?

"Tell you what, why don't you both come round?" I say to Christine and Torben. "Seven p.m. I won't take no for an answer."

9

Chemistry Stinks

I GET OFF the bus two stops early because I can't stand the smell anymore. At first I didn't want to believe that the stupefying body odor was coming from the (visually) attractive man sitting nearby, but as I left I clearly identified why he smelled so bad: TMHA, trans-3-Methyl-2-hexenoic acid. TMHA is related to *caproic acid*, a *fatty acid* named after the Latin word for goat, *capra*, because of its strong goaty smell.

Caproic acid is a *saturated fatty acid*, which means that the carbon chain contains only single bonds, no double bonds. If we added a double bond to caproic acid, we would get an *unsaturated fatty acid*. If we then added a methyl group to the double bond, we would get the wonderful TMHA molecule, with its typically goaty and literally stunning smell of sweat:

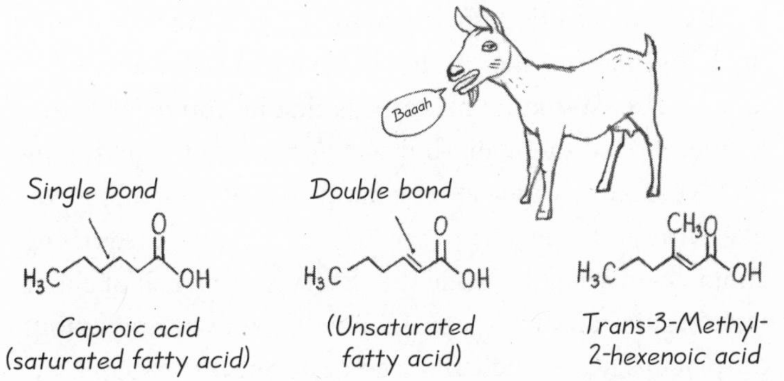

Single bond

H_3C ⌇ O OH

Caproic acid
(saturated fatty acid)

Double bond

H_3C ⌇ O OH

(Unsaturated
fatty acid)

CH_3 O

H_3C ⌇ OH

Trans-3-Methyl-
2-hexenoic acid

Now, you might find this disgusting and prefer to focus on saturated and unsaturated fatty acids, but we'll come back to them over dinner. For now, let's talk a little more about stinky molecules that may not be fun to smell, but are fascinating all the same.

Odors are caused by *volatile* molecules, which vaporize easily. When I smell something nasty, it's only because these nasty molecules have physically gotten into my nose. Yes, while I was on the bus, a bit of sweat actually flew out of that guy's armpits and into my nostrils. Sometimes reality is hard to take, I know.

ORGANIC CHEMISTRY IS associated with intense odors. The most wonderful aromatic and flavoring substances are organic molecules, but so are the nastiest smells. When you study chemistry, organic chemistry is abbreviated to oc (among German chemists at least), two letters that cause delight in some people and horror in others. There's a lot to memorize when you start out in oc. I remember sitting at home one day, doodling structural organic formulas, and my housemate saying, "I can't believe you know what they all look like." What he meant was that he couldn't believe I'd memorized them all—but what's actually fascinating is that we have any idea at all what chemical structures look like, without being able to see them. This is the great thing about chemistry. Alongside the theory, around half of chemistry studies are made up of lab work, known as *practicals*. OC practicals are tough and lead quite a few chemistry students to question not just their choice of study subject, but even the meaning of life. They're also fascinating: above all else, organic chemistry is about "cooking," as chemists call it—*synthesis*, the production of new molecules from scratch. It's a pretty cool feeling to cook up molecules with your own hands that you can't see with the naked eye or even with the world's best microscope. It makes you feel a bit like a magician. OC practicals are really demanding too. And they have a very particular odor. Take any chemist to an unfamiliar university, put them in the organic chemistry building, and they'll find the OC practical rooms by smell alone. And it isn't a nice smell. I always felt embarrassed when I took the bus home after a day of OC practicals and made the whole place stink. That guy from earlier would have paled in comparison. The TMHA sweat molecule is just an organic molecule, and we all know that it isn't the only stinky molecule to come out of humans. Here's a small selection:

BODY ODORS

HALITOSIS
1. Methanethiol (sulfur, garlic)
2. Hydrogen sulfide (sulfur, rotten eggs)
3. Dimethyl sulfide (cabbage, sulfur, sweetish)

UNDERARM SWEAT
1. Methylhexenoic acid (goat)
2. Methylsulfanylhexanol (onion)
3. Hydroxymethylhexanoic acid (cumin)

FARTS
1. Hydrogen sulfide
2. Methanethiol
3. Dimethyl sulfide

SMELLY FEET
1. Methanethiol
2. Propionic acid (sharp, rancid, sour)
3. Isovaleric acid (cheesy, fermented, rancid)

Before you think "Ha, that's funny" and read on without another thought, take a second to consider how we know all this. Do you think the chemical composition of a fart was deduced through theory alone? Oh no, there were experiments.

One study was particularly entertaining, so obviously I'm going to tell you about it: In 1998, scientists from Minneapolis studied the flatulence of sixteen men and women to identify the odor substances. At first, pretty much all the participants had to do was fart into a tube. Of course, you can't leave anything to chance in a scientific study. So the evening before and the morning of the study, the participants' food was supplemented with 200 grams (7 ounces) of beans and 15 grams (½ ounce) of lactulose, a sugar with a prebiotic effect that is broken down by the intestinal bacteria, thereby forming gas.

The remarkable thing about the study was how it was analyzed. Alongside common methods such as gas chromatography, two "judges" were employed to assess how unpleasant the odors actually were. Why only two? Well, you try finding people willing to smell fart samples in the name of science—and, of course, you need a very sensitive nose to assess them with the greatest possible scientific accuracy. In any case, these two judges had previously proven to have very sensitive noses and the ability to assess both odor quantity (how strong is the odor?) and quality (how do the odors differ?) particularly well. They rated various odor samples on a scale of 0 (odorless) to 8 (very offensive). They also had to precisely describe the smell of individual, isolated gases. Sulfurous? Rotten? Sweet? A simple "disgusting" wouldn't be specific enough.

A fart is largely made up of odorless gases like hydrogen, nitrogen, and carbon dioxide (CO_2). The researchers discovered that hydrogen sulfide is the most prolific of the stinky gases (smells like rotten eggs), followed by methanethiol (spoiled vegetables), and dimethyl sulfide (a very unpleasant sweet smell).

So what do we do with this knowledge? Well, not every piece of research has to have a practical purpose; it's valid even if it doesn't answer the "What's in it for me?" question. Science's main aim is to better understand the world, and that includes flatulence. This study went even further, and this is the really funny part.

In another experiment, the researchers had the participants wear pants made from airtight material and sealed them with duct tape on the thighs and hips. To be totally sure that no gas could escape, the participants were briefly placed in water—like a patched bicycle tire—to check that gas didn't come bubbling out of their pants. They were "wired up" via hoses to control the release (and prevent the loss) of gas from the pants so that it could be analyzed. The pants contained a foam cushion with a layer of activated charcoal on its surface to absorb molecules containing sulfur. So not a whoopee cushion, but an anti-whoopee cushion! This allowed the researchers to test how many stinky molecules can be intercepted by an activated charcoal device placed inside the pants and the extent to which this alleviates the smell. There were even placebo cushions in which the activated charcoal was sealed under an airtight layer of plastic. The methods were all correct.

What was the outcome? Well, the pants with activated charcoal cushions definitely emitted fewer smells. The anti-whoopee cushion absorbed more than 90 percent of the sulfurous gases. I should add, however—and I think this is the best bit—that the cushion measured $43.5 \times 38 \times 2.5$ centimeters ($17 \times 15 \times 1$ inches), like a huge pillow! If the only way to reduce flatulence is to put a massive cushion down your pants, then it's no wonder these products haven't made it onto the market. That's research in a nutshell. There can be a huge gulf between promising results and practical applicability.

Luckily, the supposedly "stinkiest molecule in the world" doesn't occur in nature. This small molecule is called thioacetone and looks pretty harmless:

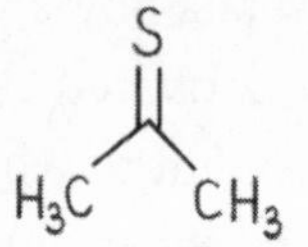

Thioacetone is difficult to extract in this form. Instead, you can produce a *trimer*, which in principle consists of three thioacetone molecules that join forces in a cyclical structure. When you heat the trimer, it splits and you can release the thioacetone:

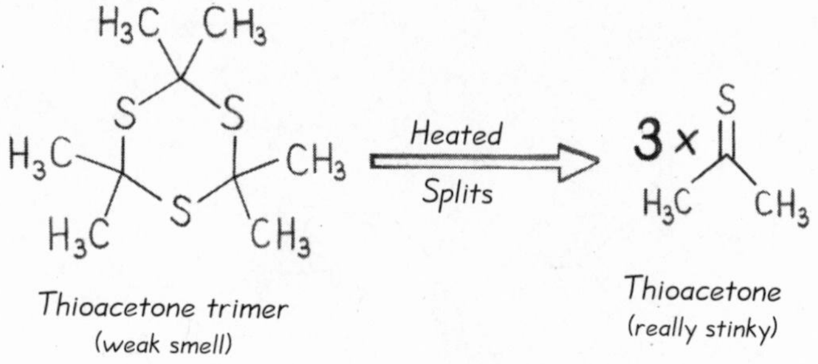

Thioacetone trimer
(weak smell)

Thioacetone
(really stinky)

But do you want to? This was first achieved by chemists in Freiburg, Germany. In 1889, they described their experiment as follows:

> As the freshly prepared reaction product [...] was carefully cooled and distilled with water vapor, the odor quickly spread to distances of ¾ kilometers, reaching far-off areas of the city. Residents of the streets neighboring the laboratory complained that the odorous substance had caused fainting, nausea, and vomiting in some people.

How do you know a true scientist? Even the nastiest stench won't curb their curiosity. Only a *"flurry of complaints"* convinced them to abandon their attempts. Initially, they didn't even expect the stinky molecule to have a specific use (maybe as a weapon?). It was simply a molecule that was extremely difficult to isolate—reason enough to give it a try! Science is all about identifying the limits of possibility.

The Freiburg scientists are early proof that organic chemists are the craziest of all chemists. At least, I think so. They're often the hardest-working too. My dad is an organic chemist by nature, even if he later switched to polymer chemistry. Happily, his doctorate focused on pleasant smells, such as freshly baked bread. My mom often recalls how nice he smelled when he came home in the evenings. My husband is also an organic chemist, but his doctorate was a lot less pleasant for me. Matthias's desk was in the lab; there was no separate office. It was the same during my PhD, but I deliberately chose a research field that was as nontoxic as possible. Matthias, meanwhile, would sit in the lab day after day with five other organic chemists who spent all their time cooking with noxious substances and sitting in their own vapors. Lab workers may have fume hoods—a sort of large, efficient exhaust hood—and good lab technology minimizes contact with chemicals, but you can't totally avoid exposure to these substances. We had a separate laundry basket just for Matthias's lab clothes. After work, he had to take them off and shower before I'd go anywhere near him. He came home almost every day smelling of OC. If little old me could smell it at home, what would Matthias have been smelling and inhaling all day? I got really angry—at his boss, at the entire institute of organic chemistry, at the whole university. How can it be that in a country like Germany, chemistry doctoral students at some institutions still don't have offices separate from their labs?

IN NATURE, A bad smell tells you to run away. The reason bodily waste smells so bad is that it may contain germs—it needs to deter us. Not everything that stinks is harmful, and not everything that's harmful stinks. That would be practical, but harmful substances aren't always obvious.

When I started my chemistry studies, I had great respect for acids. One of a student's first experiments is *acid-base titration*, which involves handling hydrochloric acid. We were all terrified of getting chemical burns. We got to grips with it all pretty quickly; as time goes on, you become a lot more sure of yourself, and your lab skills improve. Eventually, you become almost grateful for acids—if you get some on your skin, you'll realize straight away and can take appropriate measures. The nastiest chemicals are the ones whose effects you don't notice, but may cause cancer years down the line.

In our first OC practical, one of the lab experiments that made the greatest impression and taught us to work cleanly involved synthesizing a dye called crystal violet. It's very pretty; you end up with gold-bronze, needle-shaped crystals with a metallic sheen. The reason for its name only becomes apparent when you dissolve these crystals in water or other polar solvents. A tiny amount is enough to give an intense, blue-violet color—which is why the lesson comes not in producing the dye, but in cleaning your lab equipment afterward. We scrubbed and scrubbed but the color only intensified. It showed us just how much a substance can spread in the fume hood or on our lab coat. Because we did this experiment at the start of our studies, we hadn't exactly perfected our lab techniques, and the violet was everywhere. Even weeks later, areas of the fume hood would suddenly turn violet. I couldn't help thinking of this later, when I worked with toxic, but colorless, substances.

I DO MISS the lab a little. I often get nostalgic when I visit Christine, which annoys her; she tells me not to romanticize it. Now that I'm walking in the sunshine, wearing my sandals, I'm glad I don't work in a lab anymore. I almost never had semester vacations because our lecture-free periods were usually filled with practicals. And I hardly took any time off during college and my PhD. So for nine years, I spent almost every summer in the lab, wearing long pants, closed shoes, and a white coat. Sometimes it was so hot that my sweat steamed up my safety glasses. I don't miss *that* at all.

HUMANS ARE DESIGNED for hot temperatures; I mean, we're very good at sweating. Aside from the stinky molecules, sweat is largely made up of water, and this water can vaporize. States of matter don't change by themselves; the water molecules, which attract and hold on to one another in their liquid form, must be separated by adding energy, for example through heating. When our sweat wants to vaporize, it can simply take this energy from our warm bodies, and that's exactly what it does. It extracts the warmth from our body that it needs to vaporize, actively cooling us down in the process.

Looking at it this way, it's pretty stupid to use anti-perspirants. Don't get me wrong, I'm absolutely in favor of deodorants—I don't like being forced off the bus because someone can't control their trans-3-Methyl-2-hexenoic acid. But deodorants and antiperspirants are two different things.

Deodorants simply combat the smell. They contain anti-bacterial substances such as triclosan. Sweat itself doesn't smell. The trans-3-Methyl-2-hexenoic acid and friends are simply the metabolic products of bacteria. Again with the bacteria! In a way, humans are nothing more than walking, talking ecosystems for bacteria. These little single-celled organisms dominate our planet, and very skillfully too, for example by residing unnoticed in our armpits. When odorless sweat leaves our pores, it's immediately guzzled up by the bacteria, which then burp out various stinky molecules. Antibacterial ingredients make life difficult for these bacteria—and paired with a bit of perfume, deodorants can make bus journeys more pleasant for all concerned.

Antiperspirants also contain aluminum salts, which *precipitate* proteins in our armpits. All this means is that the aluminum salts form tiny plugs that clog our sweat pores to stop the sweat coming out. Not the most elegant solution, to my mind. Think about it: clogged sweat pores. Not exactly pleasant, right?

Anyhow, aluminum salts have become notorious because high doses of aluminum are toxic. When people became aware of this, everyone started selling deodorants without aluminum, which added to the fear. However, the newest data from the European Commission show that the amount of aluminum that actually enters the body via cosmetics and deodorants is negligibly small. So antiperspirants are harmless, but I still find the thought of clogged sweat pores a bit disturbing, even without the breast cancer. And yet I use antiperspirants on a regular basis because sweat marks are embarrassing. So I have a question for humanity: Why can't we, as a society, just accept sweat marks?

There's a panting dog in front of me, being taken for a walk. The poor thing doesn't even have sweat pores. My thoughts then turn to kangaroos. The only way they can stand the midday heat in the Australian desert is to lick their fur; their saliva cools them as it evaporates. Dogs and kangaroos would laugh us out of town if they knew that we voluntarily block our sweat pores.

Hannes, my physicist friend, always wears functional clothing in the summer to help the sweat evaporate. This seems sensible at first, but it's also pretty egotistical—perhaps because he views problems from a purely physical perspective. Polyester fibers encourage the reproduction of micrococcus bacteria, which are just as bad as the bacteria in our armpits. That's why our sports clothes always smell so vile.

What we really need is for someone to invent some kind of nose deodorant. If I get off the bus because I can't stand the smell of sweat, then that's down to *me*, not the stinky guy. It would need to be a spray for our noses that turned horrible stenches into pleasant scents. Then we could spend the summer sweating ourselves cool and nobody would be bothered by the smell.

This would definitely be feasible, in theory. Room sprays contain *cyclodextrins*, cage-like molecules that can literally catch bad smells—for once, the TV ads aren't that far from the truth. A few cyclodextrins in your nostrils would catch the odorants but would also block out the nice smells, and that would be a real shame.

Our sense of smell is essential, particularly for eating. Taste sensations come not just from the food that touches our taste buds, but also from volatile aroma molecules that go from our food to our noses. If you hold your nose, an apple and an onion taste remarkably similar.

WHILE WE'RE ON the subject of food, I'm planning to cook for Christine and Dino, but I still need chocolate for my dessert. Want to go to the supermarket?

10

There's Something in the Water

WHAT DO YOU think of when you hear the word "chemist"?

One of the first images that come to my mind is my dad, standing in the supermarket, reading a product's list of ingredients. My dad can spend ages browsing supermarkets. Watching him as a child, I always thought it must be super cool to understand the world so well that you can read it like a book.

I'm not quite as bad as my dad, but I do keep my eyes peeled when I go shopping, particularly for marketing scams cheeky enough to exploit the public's general ignorance of chemistry. When I enter the supermarket, I first have to walk through the beverage section. I spot a shelf of Smartwater, a mineral water brand from Coca-Cola that used to vex me when I lived in the United States. Now, this "lifestyle water" has made it into German supermarkets too. Smartwater is no ordinary mineral water; it's distilled water (pure H_2O) with added minerals. Ultimately, it's pretty much the same as all the other mineral waters in the supermarket and my tap water at home; it's just unnecessarily complex to produce. I have to concede that Smartwater has turned something as humdrum as distilled water into a marketing coup. "Inspired by clouds." Aww, how lovely.

And they're right—distillation really does follow the same principle as clouds: first the water vaporizes, then it condenses again somewhere cooler. In the case of Smartwater, this isn't in the sky, but somewhere far less romantic. In practice, condensation is accelerated using a cooler. The water returns to liquid form shortly after vaporizing and is collected in a container.

This water is now very pure; it leaves behind all possible impurities when it vaporizes and condenses again as pure H_2O. This seems convincing at first; who doesn't want to drink the purest possible water? Then again, our tap water is treated and purified before we drink it. Plus, water leaves behind something very important when it's distilled—minerals (that is, salts). These have to be added back into the distilled water to transform it into drinkable mineral water. Drinking distilled water isn't life-threatening (as some people claim) if you follow a normal diet; it just doesn't taste good. So you can see how laborious this all is. You could argue that this elaborate production process allows you to control precisely which salts end up in the water. That's true, but I don't know why that should be relevant for healthy adults. We get most of our minerals from our food.

So, depending on how you look at it, Smartwater is either a meaningless waste of resources or incredibly smart marketing. Even worse, it apparently uses spring water—drinking water—that's perfectly fine as it is. Spring water is full of minerals and has been filtered through layers of rock, but it's no "cloud water," is it? And this idiocy is billed as a "unique selling point." Well it's definitely unique, I'll give it that.

DRINKING WATER SEEMS to inspire inexhaustible creativity. There are online shops selling bottles of "moon water" (filled when the moon is full). A bottle of that costs the same as a cheap bottle of wine, but then, it contains (apparently important) lunar energy. And of course you can also get sun water, which is filled in the sunshine—for warmer, sunny bio-energies, obviously. Then there are gemstones that we should apparently be putting in our water. The desire for higher water quality isn't just esoteric. How many water filters do people buy to make ordinary tap water palatable? In Germany, tap water is subject to more stringent quality criteria than the mineral water in supermarkets. Stiftung Warentest, a German consumer organization, states that some expensive water brands offer lower quality than tap water. So if you buy still water from the supermarket because you like the taste of a particular brand, then go ahead, treat yourself. Aside from personal taste, in Germany—and many other countries— there's seldom a reason to turn your back on tap water.

Again and again, I'm struck by how much thought people put into their water without recognizing what it is that makes this substance so fascinating. Water is a magic molecule, even without gemstones or the light of the moon. I'd now like to explore one particular subject in a bit more depth to celebrate this wonderful molecule that has given us so much.

In Chapter 8, I described the bond within a water molecule as a polar covalent bond. The oxygen atom has a negative partial charge, and the hydrogen atoms have a positive partial charge. The water molecule is also angled, creating a *dipole* with a positive pole and a negative pole.

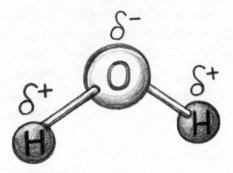

Dipolarity in a water molecule

Negative and positive charges attract one another, creating one very important property of water molecules: not only are there chemical bonds between the atoms within a molecule, there are also relatively strong interactions between different water molecules. The forces of attraction between the positive and negative partial charges are not as strong as in ionic bonds, but they are substantial enough to earn the name "bond," namely *hydrogen bonds*.

Hydrogen bonds aren't just found in water molecules. They can develop wherever hydrogen has a covalent bond with an electronegative bonding partner. The hydrogen bonds in water are particularly worth a closer look.

Without hydrogen bonds, we wouldn't exist. There would be no life on Earth. Without hydrogen bonds, water wouldn't take liquid form in the pressure and temperature conditions on this planet—it would be a gas. We can see this in molecules of similar sizes to water that can't form hydrogen bonds between one another, such as methane (CH_4) and carbon dioxide (CO_2), both of which are gases in Earth's conditions.

Water doesn't begin to boil and turn to gas until the atmospheric pressure reaches 100°C (212°F), and it's all thanks to hydrogen bonds, which help the water molecules hold on to one another.

Fish also benefit from hydrogen bonds; even in the harshest winters, ponds and lakes rarely freeze right down to the bottom. It's something to do with density. As we all know, ice floats on water. And although we see this all the time—whenever we put ice cubes in a drink, for example—we're never as surprised as we should be.

Think back to the particle model in Chapter 1. The states of matter—solid, liquid, and gas—are defined by particle density. The particles in a solid are packed most densely; the particles in a liquid have a little more freedom to move and are therefore less dense; a gas has the lowest density. So the states of matter can be altered either through pressure or temperature. If you increase the pressure, you press the particles closer together, increasing the density. This converts a gas to a liquid and a liquid to a solid. If you lower the temperature, you reduce the movement of the particles, meaning that they need less space, also increasing the density. So when water vapor is cooled, it becomes a liquid (water) and finally a solid (ice).

But wait! If ice (solid H_2O) floats on water (liquid H_2O), that means that ice has a lower density than water—and that's pretty incredible! How can it be that the liquid is denser than the solid? Yep, you've guessed it; it's all down to the hydrogen bonds. This curious fact is known as the *density anomaly of water*. When you cool water down, it acts totally normally at first; just as it should, the density increases as the temperature drops. The particles get slower and slower, allowing better hydrogen bond formation and pulling the particles closer together. Water reaches its densest point at 4°C (39.2°F). Then it gets weird; if you cool it down to 0°C (32°F), the density increases again, meaning that the water molecules move farther apart.

Why do they do that? Because the particles are moving so slowly that the water molecules now have the time and peace they need to arrange themselves properly—they begin to arrange themselves symmetrically until they form an ice crystal lattice. When you look at a snowflake or ice crystal, you can virtually see this ordered structure. The symmetrical patterns in a snowflake are simply the result of the symmetrical arrangement of the atoms inside it. In an ice crystal, every oxygen atom is surrounded by four hydrogen atoms. Two of these have a covalent bond, while the other two are connected by hydrogen bonds. This lattice structure has relatively large hollow spaces and therefore a lower density.

Why is this so important for fish in a lake?

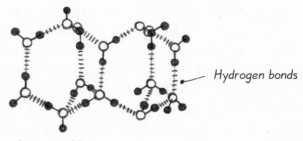

Hydrogen bonds

Ice crystal lattice

When the water in the lake grows colder in the winter, it sinks to the bottom (the colder, the denser, or "heavier," as we tend to say). Because water is most dense at 4°C (39.2°F), the water at the bottom has a temperature of 4°C (39.2°F), gradually growing colder toward the surface. At some point, the surface of the lake begins to freeze—from top to bottom. If water didn't have this density anomaly, ice would be heavier than water and the lake would freeze from bottom to top. If ice formed from the bottom to the top and the cold winter air were added from above, the lake would be far more likely to freeze all the way through. Because lakes freeze from top to bottom, the layer of ice acts like an insulator for the layers of water below it, and the fish have liquid water to swim around in—and breathe in—all winter long.

The density anomaly of water also allows humans to enjoy the winter—ice-skating wouldn't really work without it. If you think about it, ice-skating is a true curiosity. Why does it only work on ice? Why can't we do it on all solids, like asphalt? Because we don't actually touch the solid ice! We hover on a thin layer of water that forms when our skate blades exert pressure on the ice. With other substances, we'd have to reduce the pressure to change the state of matter from solid to liquid. The water anomaly makes it possible; increased pressure presses the particles closer together, the ice gives up its wide-meshed lattice so they can move closer, and a liquid layer forms that we can glide across effortlessly. Ants wearing miniature skates wouldn't be able to ice skate, because they would be too light and wouldn't exert enough pressure to create the necessary layer of water. For them, it really would be like skating on asphalt.

Some insects can walk on water—water striders even get their name from it! Again, it's all thanks to hydrogen bonds. The internal cohesion between the water molecules gives liquid water a relatively high *surface tension*, which we discussed in Chapter 3 with our soap bubbles. It's a bit like individual pieces of wood that have to be bound together to make a raft. If every water molecule were simply swimming around on

its own, the water strider would sink immediately, but because the molecules "hold on to" each other via hydrogen bonds (forming a sort of raft), a fine cross-linking develops that can hold a water strider.

A great way to see this in action at home is to put a paper clip in a glass of water. If you place the paper clip carefully on the surface of the water, it floats.

HOME EXPERIMENT NO. 3

YOU NEED:
- 1 glass of water
- 1 paper clip
- Dish soap

GENTLE CLEAN

1. The paper clip floats

2. Add a drop of dish soap

3. The paper clip sinks

And that's despite the paper clip being made of metal and having a greater density (or being "heavier") than water. The paper clip shouldn't be able to float at all, but the water's surface tension carries it. If you decrease the tension by adding a drop of dish soap to the water, releasing a few surfactants on the surface, the raft "softens" and plop! The paper clip immediately sinks.

Above all else, water is an important *solvent*. The substances we need to live—like salt and nutrients—are dissolved in water. Even we are largely made up of water, and all the metabolic reactions in our body take place in a watery solution. Our kidneys, our body's garbage filters, use water to flush out waste in the form of urine. When not being used as a solvent or means of transport, water is incorporated and converted into other substances as a chemically active reaction partner. And we've already discussed its role as a coolant—sweat—in Chapter 9.

Even that doesn't seem enough to keep us fascinated. Not far from the Smartwater, I find a shelf of "oxygenated water"—mineral water enriched with extra oxygen. Apparently, it's recommended for athletes. Seems plausible at first; the oxygen levels in our blood play a key role in athletic performance, which is why EPO (short for erythropoietin) is such a popular doping agent among endurance athletes. It increases the number of red blood cells, and the more red blood cells you have, the more oxygen can be transported through the blood to the muscles. So is oxygenated water a kind of mild performance enhancer?

Luckily (or unfortunately?), no. First of all, even if you inhale pure oxygen, you can only increase your maximum oxygen intake by 5–10 percent; our blood can't take in much more oxygen than that (unless you're taking EPO, of course). You shouldn't inhale pure oxygen anyway; it becomes pretty dangerous after an hour or two at most, because oxygen contains a small amount of nasty, reactive oxygen radicals that can attack the lungs. So why not drink dissolved oxygen instead?

Then we come to the second problem—it isn't all that easy to dissolve oxygen in water. How well a gas dissolves in water depends mainly on the pressure. Gas dissolves better in water under high pressure, which is why bottles of sparkling water are filled under pressure (to dissolve as much

CO_2 in the water as possible). When you open a new bottle of sparkling water, you'll notice that the pressure drops suddenly and a whole load of CO_2 gas escapes immediately. The same thing happens with dissolved oxygen; it's just that oxygen dissolves much more poorly than CO_2. A single breath of fresh air provides as much oxygen as a whole liter (a quarter of a gallon) of water enriched with oxygen.

The final problem is that our digestive system isn't built for gas exchange. It makes far more sense to get your oxygen supply via your lungs; that's their job. When we inhale oxygen, the gas leaves our lungs and enters the bloodstream. Not the case for our stomachs and intestines. Only a tiny proportion of gases ingested in drinks make it into the blood; the rest leaves our body in uncouth belches—but if you want to enrich your burps with oxygen gas, then I recommend oxygenated water.

Experiments have been conducted on all these matters. Does drinking oxygenated water measurably boost performance? No evidence has been found, but we probably shouldn't ignore the placebo effect. Simply believing that a particular type of water makes you fitter can itself make you fitter. Forgive me for pointing this out, but now that you've read this, the placebo effect shouldn't hold any sway over you whatsoever. This means you don't need to spend your money on weird water products that are nothing more than marketing gimmicks.

THERE ARE MANY myths surrounding water. One particularly stubborn fear is that sparkling water (containing carbonic acid) can be harmful. As the name suggests, carbonic acid is an acid. While the pH value of still water is slightly below the neutral pH level of 7, sparkling water has a lower pH value of up to 5. On the one hand, this has a slight antibacterial effect (remember the acidic preservatives in Chapter 7?). Micro-organisms find it harder to multiply when the pH level is acidic. Our digestive system isn't bothered by a pH value of 5; after all, we consume lots of acidic foods every day. Fruit, coffee, chocolate, and milk products contain acids. By the time they get to the stomach, these foods will have encountered gastric acid, which has a pH value of 1 and is so acidic that even the most carbonated of waters don't impress it. Looking at it this way, it really doesn't matter whether the water we drink is a bit more acidic, particularly when it comes to carbonic acid; once the CO_2 is gone—even if just through a decent belch—the acid is gone too. Carbonic acid is nothing more than CO_2 dissolved in water.

In 2017, however, Palestinian researchers made headlines by claiming that carbonic acid (mineral water containing CO_2) makes you hungry. Apparently, the pressure in the stomach activates ghrelin, the "hunger hormone." Germans are obsessed with sparkling water—it's our default drinking water—so the news hit them particularly hard. The study alone didn't convince me. First, the tests were only run on rats; second, lots of other hormones and factors are involved in regulating the appetite. It's interesting as a first indication, but far from proves that sparkling water actually stimulates the appetite.

Still water is definitely the best option for some people. What makes sparkling water so refreshing isn't just the tangy, tart taste, but also the CO_2 bubbles that pop against the palate. The CO_2 gas creates unrest in the stomach and increases belching. So if you suffer from acid reflux, frequent flatulence, or a gastric emptying disorder, it's better not to fill your stomach with more gases than necessary.

Ultimately, whether still or sparkling, the main thing is that you like the taste and drink plenty of water. If energized moon water stops you from drinking loads of sugary soda, then be my guest. The only thing you're damaging is your bank balance. As I said, I recommend good old-fashioned tap water; that's what I drink at home.

I LEAVE THE beverage aisle and head for the candy section. Sweet things are amazing. I wouldn't want to go without them, but I much prefer eating chocolate to drinking cola (and feel a lot less guilty about it). Sugary drinks are "empty calories," which is what makes them so insidious. You con-sume calories without getting any nutrients and don't feel particularly full afterward. Most dangerous are probably the pre-made smoothies now available in every supermarket; it's easy to feel like you're doing something good for your body, which means you might drink more. Most smoothies contain just as much sugar as cola, sometimes even more. Take a look at the ingredients next time you go shopping. The reference value for cola is around 11 grams of sugar per 100 milliliters (this means 1.4 ounces of sugar in a standard 12-ounce can of cola).

Smoothies seem healthy because they are made from "100 percent real fruit." Unfortunately, this isn't the same as actual fruit. To make smoothies nice and smooth, manufac-turers often leave out the fruit skins and add in juices. This means the sugar concentration is higher than in normal fruit, which is full of fiber. It's easy to drink large quantities of bottled smoothies, but if you tried to eat the equivalent amount of fruit, you'd get full before you could finish.

What's especially interesting is that even if you blend whole fruits at home to make a drink, it's less satisfying than eating them—and that's according to science. The consistency of our food alone can affect how full we feel, and liquid food doesn't fill you up as much as solids. So I grab three bars of chocolate with a clear conscience and make my way to the checkout.

SHOPPING LIST

- 230 g dark chocolate
- 120 g butter
- 50 g flour
- 4 medium-sized eggs
- 80 g sugar
- 1 tsp vanilla extract
- 1 pinch salt
- 1–2 tsp strong-brewed espresso

11

Culinary Therapy

THE DOORBELL RINGS and I jump. I haven't ordered anything, have I? Christine and Dino won't be here for another hour—probably more, since Christine's always late to private gatherings. For a few seconds, I freeze, paralyzed with fear, until I remember that I'm an adult and go to the door.

This fear of the doorbell goes back to my early student days. My first housemate gave me a vital piece of advice: never—and I mean NEVER—open the door! Friends never dropped by unannounced, we had no contact with the neighbors, and delivery services like Amazon didn't exist yet. So if the doorbell rang, it could only be bad news—either Jehovah's Witnesses or, even worse, the TV licensing people!

If I'd known back then that I'd end up working for a public broadcaster…

I open the door and am astonished to see Christine.

"I can't do any more work today," she says.

"Culinary therapy?" I ask.

"Culinary therapy!" Christine shouts, flinging her arms in the air.

WHEN WE WERE frustrated by our PhD research—which was often—Christine would come round to my place and we'd soothe our souls by cooking a three-course meal. Culinary therapy. It's been a long time since we've cooked together, and I'm so delighted that I pour Christine a glass of wine right away. This calls for a celebration.

"Where's Dino?" I ask.

"He'll be along later. One of his reactions was still running." Christine takes a sip, clearly feeling badly that Dino's working while she's drinking wine. This is the first evening in about a month that she's left the lab before 8 p.m. Research requires a high frustration threshold. Sometimes, you'll work for weeks or even months to confirm a particular hypothesis, only to one day run a minor experiment and discover that the whole concept was crap and your work was all in vain. Data can be tough. We all make mistakes, and we're constantly being reminded of it—but when your own failure is presented to you in numbers and data, it's especially humiliating. (I like to tell myself that it's character-building, but sadly people like King K disprove that theory.) Research also takes an annoyingly long time, demanding a tricky blend of patience and resolve. It can take years to see any notable progress, despite the blood, sweat, and tears. This is why Christine and I find cooking so satisfying: it's very similar to lab work, but the end product appears very quickly. And you can eat it. It doesn't get better than that.

"WHAT'S THE PLAN?" Christine asks.

"Fondant au chocolat for dessert," I say, "but I haven't decided on the rest. Have a look in the fridge."

We don't usually work to a recipe, and we love it. While everything has to be measured down to the microgram in the lab, in the kitchen you can follow your instincts and it still works out. Usually. This can be really liberating for chemists. Of course, a basic understanding of chemistry helps to make sure you don't mess it up completely. I've associated chemistry with cooking and good food since my childhood.

Cooking is one thing, but baking is a totally different story, chemically speaking. When people say that they like to cook but not to bake, it might be because baking is harder to improvise. Baking really is pure chemistry. You need to have some experience or knowledge to bake without a recipe and ensure it turns out OK. When you cook, you might add too many spices or cook the food for too long or not enough. Mistakes can be more fatal in baking. The cake might collapse completely, or the cookies might melt together on the baking sheet. Christine and I know the basic rules but tend to bake to existing recipes that we adapt over time.

While we prepare our feast, I'd like to use our fondant au chocolat to show you the fascinating chemistry of baking. This information will help with all your future baking. And I'll throw in my recipe for good measure.

Fondant au chocolat is a small, warm chocolate cake with a gooey filling. I think it's one of the dreamiest desserts out there, and it's so simple to make. The recipe begins with one of the most heavenly things on Earth: chocolate! Use 230 grams (8 ounces), ideally of dark chocolate that's 45–60 percent cocoa. I prefer to have more cocoa. Cocoa contains some fascinating molecules, including *theobromine*, which looks almost exactly like caffeine:

Theobromine   *Caffeine*

And it works almost exactly like caffeine too (think back to Chapter 7, where we looked at adenosine and its receptor "parking spaces"). Theobromine also competes with adenosine for space in the receptor. Now, chocolate might make some people so happy that they're practically euphoric, but despite the uncanny similarity in chemical structure, theobromine is much less efficient than caffeine at keeping us awake. For one thing, theobromine doesn't fit into the parking space as well as caffeine and can't force out the adenosine molecules quite so aggressively. There's no need to worry about chocolate keeping you awake at night—we don't consume enough theobromine for that.

Just like caffeine, however, theobromine does become poisonous at a certain quantity (the dose makes the poison). Thankfully, we'd have to scoff impossible quantities of chocolate to be in any danger of overdosing on theobromine. We'd either puke or give up in disgust long before that happened. It's far more dangerous for dogs, though. The lethal dose of theobromine is much smaller for them, mainly because it takes them a lot longer to metabolize it. While our bodies quickly convert this potentially poisonous stimulant into other, harmless molecules, a dog's theobromine chemistry isn't as quick, causing the molecule to build up in its body. This leads to palpitations, muscle spasms, nausea, vomiting—and even death. So if you're enjoying some chocolate and your dog looks at you with those big, sad eyes, don't give in!

It's common knowledge that dogs shouldn't eat chocolate; the same goes for cats too. However, cats have an advantage over dogs and most other mammals: they can't taste sweetness. Cats' taste buds have such different proteins that they can't detect sugar or carbohydrates, meaning that the corresponding signal isn't sent to the brain. So a cat is less likely to stare jealously at your chocolate because it has no idea how blissfully sweet it is. For safety's sake, I'd still hide your chocolate from your cats though—as all those YouTube videos show, cats are endlessly curious.

We humans should be thankful that we can tolerate theobromine so well. Although chocolate is made up largely of sugar and fat, it's the combination of these two ingredients and cocoa that makes it so irresistible.

FIRST, I MELT the chocolate in a bain-marie. Adept bakers can melt chocolate in the microwave too, but there's little more satisfying than stirring chocolate and watching it melt (and the smell is amazing!). The advantage of a bain-marie is that, regardless of the stovetop temperature, the water can't exceed its boiling temperature of 100°C (212°F). This prevents the chocolate from overheating, which would turn it all horrible and lumpy.

When you overheat chocolate, the first thing you notice is that its ingredients (sugar and fat) actually want nothing to do with one another. Sugar is a hydrophilic, polar substance, while fat is a hydrophobic, nonpolar substance. Lecithins, a class of surfactants typically extracted from soybeans, ensure that they blend together evenly. Just like the surfactants in our shampoo, lecithins are amphiphilic molecules that act as *emulsifiers*. They sit on the boundaries between sugar and fat and stabilize the mixture. When we overheat chocolate, the lecithins can't do their job anymore, and that's that—we're stuck with lumps of cocoa fat, milk fat, sugar, and cocoa particles.

The risky thing about a bain-marie is that the water mustn't get into the melting chocolate. The best way to avoid this is to make sure the water doesn't actually boil; then it won't bubble and water drops won't accidentally end up in the chocolate. A warm bain-marie is absolutely fine; chocolate melts in the mouth, so it doesn't need particularly high temperatures. If water gets into the melting chocolate, then we'll have a very hydrophilic substance in the mix. Water and sugar stick together straight away; even the tiniest amount of water in a bag of sugar will turn it all clumpy. Something similar happens with chocolate; tiny amounts of water create lumps that are hard to remove.

THE CHOCOLATE SHOULDN'T be melted alone. I add 120 grams (4.2 ounces) of butter—lots of lovely, lovely fat (what can I say, I'm lipophilic!). I'm quite interested by the way we talk about fats and oils in everyday life. "Saturated fatty acids," "unsaturated fatty acids," "trans fatty acids," "omega-3 fatty acids"—it's rare to hear so many chemistry terms in common parlance. It's great, but I fear that, despite their frequent use, it's not entirely clear what these various terms actually mean. So how about a little look at the chemistry of fats?

Saponification has already shown us that fats and oils are made up of *triglycerides*, a fusion of three *fatty acids*. Fatty acids are long-chain molecules; long chains of carbon (C) atoms, to be more precise. Every C–C bond contains energy that the body can use for its metabolic processes. Fats have the most energy of all nutrients. "Fantastic!" our body thinks, stuck in the hunter-gatherer mindset. Our love for this valuable energy source ensures that we'll devour it whenever we can. Unfortunately, fat is available everywhere these days, and a valuable energy source has become an unhealthy luxury.

Not all fats are equal; unsaturated fatty acids are good, while saturated fatty acids are bad! At least that's what people say. Is this true, and what exactly is the difference between unsaturated and saturated fatty acids?

So: every carbon atom can form four bonds. Within a fatty acid chain, every C atom is bonded to two more C atoms, meaning it still has two bonds available. If you add two hydrogen (H) atoms to every C atom in the chain, you get a *saturated fatty acid*—saturated with hydrogen.

In contrast, *unsaturated fatty acids* contain C=C double bonds. For every double bond, two H atoms have to be thrown out. So there's only one H atom attached to every double-bonded C atom—making it "unsaturated."

We also distinguish between *monounsaturated and polyunsaturated fatty acids*. This relates to the number of double bonds. One double bond means monounsaturated, and more than one double bond means polyunsaturated.

You may have realized already that the next few paragraphs will take some concentration; there are some pretty tricky terms in the chemistry of fats. It's very easy to confuse "saturated" and "unsaturated" when reading quickly; even I do it.

Plus, we automatically associate "un," like in "unsaturated," with something missing. And, yes, the H atoms are missing from unsaturated fatty acids, but this isn't all that important. In unsaturated fatty acids, everything revolves around the presence of double bonds. Not literally, though—nothing can actually revolve around a double bond.

To see what I mean, get hold of some cherry tomatoes and toothpicks. Two cherry tomatoes connected by one toothpick represent a single bond. You can easily rotate both tomatoes in opposite directions, indicating that a single bond can move freely. If you connect the two tomatoes with two parallel toothpicks, you end up with a model of a double bond. This bond is rigid; you can't rotate the tomatoes without destroying them.

HOME EXPERIMENT NO. 4

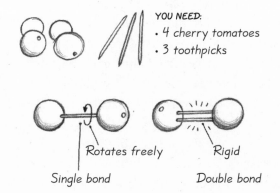

YOU NEED:
- 4 cherry tomatoes
- 3 toothpicks

Rotates freely — Single bond

Rigid — Double bond

This means that all double bonds are rigid. In unsaturated fatty acids, they are usually "kinked":

Palmitic acid

Saturated fatty acids

Oleic acid

Unsaturated fatty acids

This simple kink in the molecular structure visibly alters a fatty acid's physical properties. Saturated fatty acids that aren't kinked tend to produce solid fats, while unsaturated fatty acids with a kinked shape tend to produce liquid fats. Picture it like this: saturated fatty acids can stack and layer themselves more easily, which also makes it easier to form a fixed structure. Unsaturated fatty acids with a kinked shape are unwieldy and harder to stack, meaning that unsaturated fats tend to be liquid oils. Therefore, the state of matter tells you whether you're dealing with unsaturated or saturated fatty acids. However, the line between solid and liquid is literally fluid because saturated and unsaturated fatty acids are often mixed—like in chocolate, but more on that in a minute.

First, it's crucial for you to know that unsaturated fatty acids don't necessarily have to be kinked. Whenever you have a double bond within a C chain, there are two possibilities—cis or trans (kinked or unkinked):

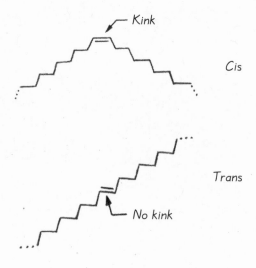

In our natural diet, almost all the fatty acids we encounter are cis fatty acids with a kinked shape. Small quantities of trans fatty acids are found in animal fats—specifically, in animals that chew their cud. For example, milk fat contains between 1 and 6 percent trans fats. So when we talk about unsaturated fatty acids in everyday life, what we actually mean is cis fatty acids. Trans fatty acids are so problematic that we call them by their actual name. Trans fatty acids are considered the unhealthiest of all fatty acids. No more than 1 percent of your total energy intake should come from trans fats.

Unfortunately, it took a relatively long time for this to become apparent. At first, nobody was bothered that hardening fat created large quantities of trans fats as a by-product, and that these were simply left in the product. To harden fat, you take unsaturated cis fatty acids and *hydrogenate* them (*hydrogenation* indicates that this is a reaction with hydrogen). Hydrogen goes to the double bonds and convinces the carbon atoms (with the aid of heat and pressure, among other things) to give up their double bonds and incorporate them into the chain instead. So, hydrogenation is the saturation of unsaturated fatty acids.

When the German chemist Wilhelm Normann developed this process in 1901, it proved extremely useful at first. Cheap plant oils could be used to produce hard fats like margarine and lard. Fats hardened in this way are also used to make soap (see Chapter 3). Something else also happens in this reaction: some cis double bonds convert to trans double bonds (some kinks are straightened out). This is why trans fats are produced during artificial fat hardening and hydrogenation. Ironically, it was long thought that trans fats produced artificially from plant fats were healthier than animal fats. Health-conscious butter lovers swapped to margarine with a heavy heart.

Then came research. It didn't happen overnight, but evidence gradually emerged not only that trans fats weren't better, but that they weren't good at all for the cardiovascular system. From hype to no-go. The World Health Organization (WHO) called for the elimination of trans fats by 2023. The warnings took effect, and trans fats developed such a poor reputation that many food manufacturers now voluntarily avoid adding them or keep them to a minimum. Today, the German Federal Institute for Risk Assessment (Bundesinstitut für Risikobewertung) states that the intake of trans fatty acids in Germany has returned to uncritical levels.

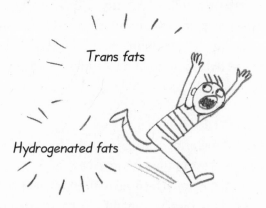

So, trans fats are bad. What about saturated and unsaturated (cis) fats?

Many studies have been conducted in this area. Nutritional studies are tricky, doomed to return contradictory results. Again, this is down to scientific methods. Thinking back to Chapter 5 and randomized controlled trials, it's immediately obvious that such clinical trials aren't generally feasible in nutritional research. It's easy to divide lab mice into test and control groups, and "blinding" is automatic when animals are used. Blind nutritional trials with human subjects are rare because—well, people know what they're putting in their mouths. Even that isn't the biggest challenge. You can precisely control what animals eat, and even how much they move, over a long time frame. Want to do the same with a group of people? Good luck!

Nevertheless, there's plenty of persuasive evidence in some areas of this subject. For example, it's recommended to replace saturated fatty acids with polyunsaturated fatty acids. The experts just can't agree whether polyunsaturated fatty acids are automatically healthier than monounsaturated fatty acids—whether "the more double bonds, the better" should be the rule for healthier, unsaturated fats.

The only fatty acids entitled to feel superior are *omega-3 fatty acids* and *omega-6 fatty acids*. "Omega-3" denotes unsaturated fatty acids with a double bond at the third C atom (hence the "3") starting from the end of the chain ("omega" means "end"). The same goes for omega-6, but with the sixth C atom from the end.

The omega-3 fatty acid alpha-linolenic acid and the omega-6 fatty acid linoleic acid are *essential fatty acids*, which means they are crucial to our survival—and our bodies can't produce them. We *have* to eat them. They can be found in various plant oils and in fish. Don't start panicking and gorging on fish and canola oil. The recommended daily amount is

250 milligrams (eight thousandths of an ounce). Anything more than that is just for fun.

FAT SHOULDN'T BE demonized or celebrated across the board. We should get no more than 30–35 percent of our total energy from fat, but at least 10 percent to ensure we get enough calories and essential fatty acids. Plus, some vitamins are very hydrophobic and get along much better with fat than with water.

As (almost) always, it's a question of balance. You shouldn't eat fondant au chocolat every day (with its 230 grams [8 ounces] of chocolate and 120 grams [4.2 ounces] of butter), but delicious food makes me happy, and that shouldn't be underestimated either. Sometimes we concentrate too much on physical health and neglect our mental health. So it's back to the chocolate!

Depending on the brand, dark chocolate contains around 30–35 percent fat. Some of this fat comes from the cocoa plant. This cocoa fat, also known as cocoa butter, is a complex blend of various fats. Cocoa butter mainly consists of three fatty acids: oleic acid (unsaturated), palmitic acid (saturated), and stearic acid (saturated). Happily, this blend remains solid at room temperature but melts at body temperature, so it melts in the mouth. Apart from the cocoa butter, chocolate is mainly milk fat. The lighter the chocolate, the more milk fat it contains. Milk fat is another complex blend of fatty acids, saturated and unsaturated, but with a lower melting temperature than cocoa fat. You may have noticed that milk chocolate is creamier than dark chocolate and goes softer when it melts.

Milk fat is also used in butter. For butter to be called butter, it must be at least 80 percent milk fat. So I add a generous helping to the fondant batter. As well as enhancing the flavor, it reduces the melting point of the mixture and makes it easier to achieve the gooey center.

Butter is also up to 16 percent water, and this is important in baking. In contrast to the water from the bain-marie, the water in the butter is already emulsified—in other words, mixed in well with the milk fat. This means that lumps don't form when it's mixed with the chocolate. Later, in the oven, the liquid water will become gaseous water vapor, its volume will increase significantly, and this will help the cake to rise. Because we often see gases as "nothing" and can't taste hollow spaces, gases aren't really seen as an ingredient. In fact, all states of matter are crucial to the overall flavor of a cake. The development of gases pretty much always plays a key role in baking. Baking powder and sodium bicarbonate are traditionally used, decomposing in the oven to form carbon dioxide (CO_2) gas and allowing the cake to rise.

WHILE THE CHOCOLATE and butter are melting in the bain-marie to form a delicious ganache, I put 50 grams (¼ cup) of flour in a small bowl and stir in a pinch of salt. Most baking recipes tell you to mix the dry ingredients before adding the ingredients that contain water, and there's a good reason for that. Flour contains a protein called *gluten*, which is sensitive to water. Gluten went relatively unnoticed for a long time, but has been attracting increasing attention—very negative attention—over the last several years. More and more people are finding that they can't tolerate gluten or that they feel better when they remove it from their diets. Science struggles to explain this—most of the people affected don't suffer from *celiac disease* (a genetic gluten intolerance) or a wheat allergy. Experts are divided and offer various explanations, including the nocebo effect.

Only one thing is certain: gluten plays a key role in baking and lives up to the "glue" in its name. Gluten actually contains two different proteins or protein classes, *gliadins* and *glutenins*. The two only come together when the flour makes contact with water, forming a three-dimensional, sticky structure: gluten. This sticky structure gives bread and noodles their elastic, chewy consistency.

So it's very important to mix the flour and water at the right time because they start to stick together immediately. Once the gluten is activated and the dough becomes sticky, it's difficult to evenly distribute other dry ingredients like sugar and baking powder—so all dry ingredients should be mixed beforehand. Bread should have a chewiness to it, but this should be avoided in cakes or fondant au chocolat, for example. That's one reason why I'm using so little flour (as well as keeping the center gooey).

THE CHOCOLATE/BUTTER MIX has melted, and I set it aside to cool. The next step is a variation on a typical fondant batter. Normally, you could mix the eggs with a bit of flour straight away and whisk the whole lot. But first I beat the (four medium-sized) eggs with a hand blender in a second large bowl until they turn foamy. Then I gradually add 80 grams (5/8 cup) of sugar. Sugar is another dry ingredient, and normally you could mix it straight in with the flour. Try it if you like—the batter will be a bit denser and less fluffy (you might prefer it this way). It's better to beat the eggs with sugar because the sugar crystals act like tiny grindstones that battle their way through the egg mixture.

Eggs contain lots of *proteins* essential for baking. Like fats, we can visualize proteins as long-chain molecules, but with a more complex chemical composition; their building blocks are *amino acids*. These chains are also much longer than those of fatty acids—so long that they tangle and fold themselves into large, three-dimensional superstructures. From the outside, a protein looks more like a sphere or some other three-dimensional shape, not like a chain.

As I made my fried egg this morning, we saw that when proteins are heated, they become solid. The heat causes the proteins to uncoil and unfold. This process is called *denaturation*. The long strands then get tangled up, forming a mesh-like structure, and the egg becomes solid. The tangled egg strands are a bit like when you get your headphones out of your bag to find the cables all tangled. (The analogy ends there, though; headphone cables are almost impossible to unravel, but with patience and a few choice expletives, they can be. But not for a cooked egg. The process of denaturing proteins can't be reversed.)

Something similar happens when you beat eggs; it's just not quite as brutal. As you physically beat them with the hand blender, the proteins partially uncoil and then start to get tangled up, a kind of "denaturation lite." If you use only the egg whites, you can beat them until they're really firm. The key is air; lots of little air bubbles get trapped, so the more stable the foam, the airier the dessert. If we were making a chocolate soufflé, firm egg whites would be essential for achieving the fluffy, cloudlike consistency. I've opted for a watered-down version of a soufflé. I definitely want some fluffiness to stop the dessert being too heavy, but I want it to be denser than a soufflé. So I don't separate the eggs before beating. This creates a softer foam; the yolk adds a significant proportion of fat into the mixture (fat again!) making it hard for the egg white cables to get tangled up.

Egg yolk is around 30 percent fat, which prevents you from "overbeating" the egg. Beaten egg whites, which are made up almost entirely of protein and water, are easily ruined if you beat them for too long. The proteins clump together and separate from the water, and then the egg white collapses. You can't really go wrong if you use the whole egg. By the end, the volume has increased significantly (hence the large bowl), and the delicate foam has a smooth, glistening, pale yellow surface.

Later, in the oven, the proteins will be completely denatured and will turn solid. The water in the egg plays the same role as the water in the butter: it allows the cake to rise. Together with the air that's been beaten in, this ensures that the outside of the cake is airy and the inside is gooey.

THE SUGAR DOESN'T just help with beating the eggs, obviously. Sugar is the foundation of all things sweet, but don't make the mistake of reducing sugar to just its sweetness. Sugar is also *hygroscopic*, which means it attracts and holds on to water (hence its use in preservatives, as we saw in Chapter 7). In a fondant au chocolat, which has a gooey center and should be eaten while hot, this doesn't play too much of a role. The less sugar a cake or cookie contains, the quicker it will dry out. So if you're trying to reduce your sugar intake by halving the sugar in your cake recipes, all you'll get in return are dry cakes.

Sugar is especially important for ice cream, particularly sorbet, which contains not only lots of sugar, but also lots of water in which the sugar dissolves. As with all soluble substances, the sugar content affects the temperature at which the solution melts and freezes. We see this phenomenon when spreading salt on roads and sidewalks in winter; salt water has a lower freezing point than pure water, a phenomenon appropriately known as *freezing point depression*. While pure water freezes at 0°C (32°F), a saline solution will still be liquid at this temperature. This is why spreading salt is such an effective way to protect us against slipping; the temperature has to drop further before the water turns to ice. Freezing point depression also works with sugar water and directly affects the consistency of ice cream and sorbet. The higher the sugar content, the more the ice melts; the lower the sugar content, the more solid it remains. If you make ice cream yourself, you can't just base the sugar content on your own preferences; you need to find a good consistency that's not too solid and not too mushy.

THE CHOCOLATE/BUTTER MIXTURE has cooled now, and I add a teaspoon of vanilla extract. The sweet vanilla flavors blend wonderfully with the bitterness of the cocoa. Vanilla is a very popular flavor. It's used in so many products that you might think there's a surplus of the stuff. "Vanilla" has even become slang for conventional, boring, or prudish. This flavor isn't as commonplace as it seems.

I bought a little bottle of organic vanilla extract ("from pure bourbon vanilla pods") a few months ago, when I was feeling indulgent. The reason it's so expensive is that vanilla plants are unbelievably arduous to cultivate. For a long time, vanilla (an orchid) could be found only in Central America, home to the *Melipona* genus of bee. This is one of the few creatures that pollinate the vanilla plant, which isn't particularly good at reproduction (so maybe "vanilla" *is* a good way to call someone prudish). Vanilla used to be a rare flavor enjoyed by a select group of people.

That changed in 1841 thanks to Edmond Albius, who was born into slavery on the French colony of Réunion, a small island near Madagascar. At the age of twelve, he discovered that vanilla can also be pollinated by hand. Réunion became a major exporter of vanilla and soon it was also being grown on Madagascar, which still produces a large proportion of natural vanilla—it has to come from Madagascar to be called "bourbon vanilla." That wouldn't be anywhere near enough to cover global demand. Around 18,000 metric tons (20,000 tons) of vanilla flavoring are produced each year, but only around 1 percent comes from real vanilla plants. Even today, the plants are cultivated in a very similar way to Albius's method; they still have to be pollinated by hand.

In the 1970s, scientists discovered that the vanillin molecule, the main flavoring agent in vanilla, can be produced in the lab. If you read the list of ingredients, you'll see that the imitation vanilla extract available in supermarkets is usually made with vanillin.

Vanillin

Demand for natural vanilla flavoring has grown in recent years, but there simply aren't enough vanilla plants on the planet. Aside from the laborious cultivation process, the plants don't have a particularly high yield. Around six hundred flowers have to be pollinated by hand to produce 1 kilogram (2.2 pounds) of vanilla pods. So if you want natural vanilla flavoring, you'll have to pay through the nose like stupid old me. The natural flavoring does have a more complex taste (real bourbon vanilla contains more than just vanillin), but imitation vanilla extract won't ruin this recipe; it'll still taste good.

THERE'S ONE FINAL secret to achieving the ultimate chocolate flavor: a shot of espresso. And before you think "Urgh, no, I hate coffee-flavored chocolate," I can assure you that you won't be able to taste it. (I don't like coffee chocolate either.) Theobromine and caffeine aren't the only things that are similar—chocolate and coffee also have similar flavors ranging from bitter/nutty to fruity. If you have any baking cocoa at home, taste a little bit on your fingertip—you'll see that it tastes a little like coffee, just not as intense. So adding one or two teaspoons of strong-brewed espresso to your chocolate dessert—whether a mousse or a cake—is almost like adding concentrated cocoa aroma, but with an interesting twist that cocoa powder doesn't offer.

NOW THAT I'VE added vanilla and espresso to the cooled chocolate mixture, I fold it into the beaten eggs. It doesn't need to have cooled to room temperature, just enough that the proteins in the eggs don't denature. Egg yolk denatures at 65°C (149°F) and egg white at 83°C (181.4°F). I don't mix it too vigorously, just enough to ensure an even distribution—I don't want to destroy too many of the air bubbles I've just created. Finally, I fold in the flour with the pinch of salt, again ensuring an even spread. Next is to pour the batter into greased soufflé molds or muffin molds. Any leftover batter can be stored in the fridge, ideally in molds so that they can go in the oven later.

I fill four molds and put them in the fridge for now. You can prepare this dessert in advance, then put it in the oven shortly before serving. The baking time is crucial to keeping the center gooey, so make sure you adjust it as needed. My soufflé molds are 7 centimeters (2.8 inches) in diameter, and I always fill them with a good 4 centimeters (1.5 inches) of batter. The best baking time for these dimensions is 15.5 minutes at 190°C (374°F), top and bottom heat—at least if the mixture is at room temperature. If it's been in the fridge overnight, then it needs 16–16.5 minutes. If you're using smaller molds, then reduce the baking time by a few minutes. Invite your friends round and get them to test your efforts!

CHRISTINE AND I are so engrossed in our cooking party that we don't notice Matthias walk in with Dino in tow. They both look a little vexed.

"Great, you're here," Christine calls happily. "We need help with the chopping!"

Matthias beckons me out of the kitchen and whispers, "Is that Dino?"

"Yes," I whisper back.

"We arrived on the doorstep at the same time and I didn't know who he was."

"We decided we needed some culinary therapy," I explain. "And we invited Dino too."

"He didn't introduce himself; he just followed me in silence." Matthias laughs.

"Torben's cool," I say. "He just needs to warm up a bit."

We go back into the kitchen and I get two glasses of wine for Matthias and Torben. This evening is going to be fun.

The Right Chemistry

WHILE WE CHOP vegetables and have a heated debate about whether it's better to floss before or after brushing your teeth, Christine receives six text messages from Jonas. He's one of those annoying people who send each tiny fragment in a separate text rather than using punctuation and paragraphs:

"Hey"

"How's it going :)"

"Still in the lab?"

"I can cook tonight"

"Want to drop by?"

"I can come get you"

Christine looks at me, irritated.

"What are you looking at *me* for?" I ask. I can't help laughing. "All this because of the damn toothpaste!"

Christine starts laughing too. Then she says, "It's not really about the toothpaste. That was just… the catalyst. The chemistry wasn't quite right." The double meaning makes us laugh again.

"I'll call him," Christine sighs, leaving the kitchen.

"THE RIGHT CHEMISTRY" is an interesting turn of phrase; it's by far the most positive reference to chemistry in colloquial speech. The chemistry of love! I don't know what nonchemists think of when they use this phrase, but love always makes me think of chemistry—and science. Is that unromantic? I don't know. I don't think a scientific view of the world dulls its magic at all.

The American physicist and Nobel laureate Richard Feynman summarized this perfectly in one of his interviews:

> I have a friend who's an artist and has sometimes taken a view that I don't agree with very well. He'll hold up a flower and say, "Look how beautiful it is," and I'll agree. Then he says, "See, I as an artist can see how beautiful this is, but you as a scientist will take this all apart and it becomes a dull thing." And I think that he's kind of nutty! […] I can appreciate the beauty of a flower. At the same time, I see much more about the flower than he sees. I could imagine the cells in there, the complicated actions inside, which also have a beauty. I mean, it's not just beauty at this dimension, one centimeter; there's also beauty in the smaller dimensions, the inner structure, also the processes.

The fact that the colors of the flower have evolved in order to attract insects to pollinate it, is interesting; it means that insects can see the color. It adds a question: Does this aesthetic sense also exist in the lower forms? Why is it aesthetic? All kinds of interesting questions, which the science knowledge only adds to the excitement, the mystery, and the awe of the flower. It only adds. I don't understand how it subtracts.

Feynman's words would have any scientist banging on the table and shouting, "Yes, exactly!!" I secretly hope that you're coming round to Feynman's way of thinking, even if you're not a scientist. Understanding things in detail only makes them more fascinating.

Plus, the beauty of science doesn't necessarily lie in unlocking the truth, but in searching for it. We have a long way to go before we'll be able to break down all the nuances of love into scientific concepts. I don't find it remotely unromantic to examine love, emotions, and interpersonal relationships from a scientific perspective.

EVEN WITHOUT CONDUCTING studies, I can confidently say that Matthias and I have the right chemistry. Maybe because we're both chemists—haha! Even after ten years together, I still get butterflies in my stomach when I hear his key in the door after an intense day of working from home or when he collects me from the train station after a stressful day of filming. I know it's cheesy, almost disgustingly so, but the trigger for that butterfly feeling is anything but romantic. It's a fight-or-flight reaction, the same mechanism I experienced this morning when Matthias's alarm clock rang.

So do I want to run away from Matthias, or punch him in the face? Neither—and if that's how you feel about your partner, then it's probably best for all concerned if you split up as soon as possible. Even if we perceive it as a positive thing, this physical stress response is part of falling in love. Love doesn't just make your heart race; it also increases your cortisol level. Having discovered this morning that cortisol is a "stress hormone" (along with adrenaline), we're going to explore its other side. You could even call cortisol a love hormone!

Now that you know this, it might help you to overcome stage fright by looking at it from a different angle. The fear of standing onstage or giving a presentation in front of strangers is a fight-or-flight reaction too. But people who enjoy being onstage don't feel a desire to run; they get butterflies. The underlying chemistry is the same.

Whatever the situation, the fight-or-flight reaction means that the thing we're about to do (like give a presentation) is extremely important and needs to take priority. Just like an encounter with a saber-tooth tiger, we have to give it our full attention. So some bodily functions, unimportant at present, will just have to wait—like digestion. The blood is diverted away from the stomach, causing that sluggish feeling that's awful when you're stressed but can be wonderful when falling in love. So when I see Matthias after a long day, my body says, "Drop everything, we can digest later—we need to give this lovely person our full attention!"

I'M VERY LUCKY to have someone to hug me after a stressful day. We all know the emotional impact of a hug. The whole

"Free Hugs" idea—wearing a sign and offering to embrace strangers—was done to death a few years ago, but the beaming faces and utter joy a hug can trigger, even from a stranger, are fascinating.

What exactly happens when we put our arms around another person? To answer this question, psychologists at Carnegie Mellon University in Pittsburgh conducted a study with around four hundred participants. Their method had three steps:

Step 1: They asked the participants about their social networks (real ones, not online) and the emotional support they received in everyday life. Did they have friends they could spend time with? Did they often feel socially excluded? Was there someone to whom they could confide their worries and fears?

Step 2: Every evening for fourteen consecutive days, they asked the participants whether they had encountered any social conflicts—and whether they had been hugged.

So far, so normal, as far as the methods go.

Then came Step 3: The participants were infected with a cold virus, quarantined, and observed! Pretty invasive for a psychological study. The results were fascinating. Social conflicts can cause stress (the bad kind, not the butterflies) and this, in turn, can weaken the immune system. So we're more likely to catch a cold when we're stressed. However, the people who stated in Step 1 that they had a strong social and emotional network exhibited a lower probability of catching a cold, regardless of how many social conflicts they had endured during the fourteen days. The people who stated in

Step 2 that they had received frequent hugs returned a similarly positive result.

So do hugs help fight colds? It would definitely be good to see more research on this, but until then I'll stick to my daily hug quota.

I'm very affectionate as a whole, and I probably get that from my mom. Maybe it's genetic or because she hugged me a lot as a kid. Whenever I visit my parents, I get plenty of hugs and kisses. When I turned twelve—and visited my mom's family in Vietnam for the first time—I had a lightbulb moment. We'd flown a long way and endured a bumpy bus ride on narrow, unpaved roads that seemed almost as long as the flight. It was getting dark, and by the time I got off the bus, I was truly exhausted. Then I was ambushed by a bunch of strangers who looked a little like me. Their shouts ranged from joy to hysteria; they cried; they wrapped their arms around me, smothered my forehead in kisses, and refused to let me go. So it seems to be a family thing. And I really like it, even if I did find it super embarrassing as a teenager.

My second lightbulb moment came during my studies, when I learned about the molecule *oxytocin*. This hormone plays a key role in childbirth and breastfeeding (it helps with muscle contractions in the womb, for example), hence its name, which means "quick birth" in ancient Greek. Oxytocin facilitates the close relationship between mothers and their children and between romantic partners (it's released during kissing) and is generally linked to social relationships and love. There's a reason why oxytocin is popularly known as the "cuddle hormone."

Aha, I thought. My mother's family probably has high oxytocin levels. Oxytocin soon showed me that, unfortunately, it's never so easy to explain the effects of hormone molecules.

APPROPRIATELY, I FIND oxytocin to be a really pretty molecule:

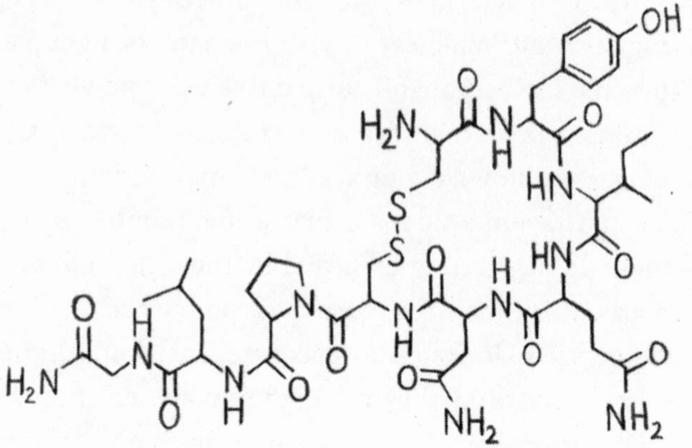

This beautiful chemical structure, and its "cuddle hormone" nickname, make oxytocin very popular among nerds and science enthusiasts. You can buy mugs, sweaters, and even necklaces featuring the oxytocin molecule. It's also a very popular subject for scientific research.

In 1979, a landmark study was conducted in which oxytocin was given to virgin rats. The hormone triggered maternal behavior, and they began to mother young rats as if they were their own offspring. In 1994, it was discovered that oxytocin plays a significant role in partner selection among prairie voles. Not only are prairie voles really cute (look them up online), they are also one of the few monogamous mammals;

they stay with the same partner until the end of their lives. What can I say, you can't help loving oxytocin.

According to a Swiss study, oxytocin boosts trust in humans. Participants were asked to play a "trust game" that involved investing money in their fellow players. The participants who'd received oxytocin first appeared to be more trusting of the other people playing the game. German researchers discovered that men under the influence of oxytocin consumed fewer calories from snacks, leading to speculation that oxytocin could stop a person from snacking when they aren't actually hungry. A cuddle hormone that fights the munchies? A potentially interesting way to treat obesity, but more research is needed.

THERE'S CERTAINLY A wide range of research, but the cuddle hormone isn't always painted in a positive light. For example, recent research indicates that oxytocin reinforces unpleasant memories. It seems to strengthen recollections of social interactions in general, be they pleasant (like a first kiss) or unpleasant (like humiliation or the fear of losing a loved one).

And while oxytocin strengthens interpersonal relationships, Dutch psychologists have observed that it could also strengthen groupthink and the exclusion of others. Just as empathy and compassion enable us to act humanely, oxytocin may make us more compassionate toward people similar to ourselves than to people we perceive as outsiders.

THE MOST RECENT research suggests that the idea of oxytocin as a cuddle hormone is probably outdated. Its influence on our social behavior and interaction is undisputed, but it encompasses both positive and negative sides. We need to look at oxytocin as follows: We are bombarded with signals, stimuli, and information in our everyday lives, so the occasional piece of social information may be lost. I might be sitting with Christine in the cafeteria but fail to notice that she's worried. Oxytocin acts like a noise filter. It ensures the release of an *inhibitory neurotransmitter* called GABA, short for gamma-aminobutyric acid, which we will explore in more detail in the next chapter. This signal inhibition reduces the noise, allowing us to be more alert to social stimuli and information.

Researchers are therefore exploring oxytocin's potential to help people who are on the autism spectrum. Most people with autism find it difficult to classify social information, such as reading a person's emotion from their facial expression. So far, studies have come unstuck due to poor reproducibility or have not been able to establish an effect with repeated application. However, some people with autism whose oxytocin levels have always been low may benefit from additional oxytocin in the future.

INTERESTINGLY, MORE AND more parallels are emerging between oxytocin and alcohol. First they were observed to have similar, externally apparent effects. Both oxytocin and alcohol can reduce fear and stress while boosting trust and generosity. They also both have the same drawbacks: aggression, risk-taking, and a tendency to show positive bias toward your own group. And there are uncanny parallels in their neurological effects too. Alcohol also boosts the inhibiting effect of the neurotransmitter GABA, although the mechanism is different (as we'll see in the next chapter). In any case, there may well be something to the expression "love drunk."

CHRISTINE IS ANYTHING but love drunk when she returns from calling Jonas. But she gets a hug from me in compensation. And I pour her a little more wine, just to be certain.

A Passion for Facts

WE SIT AROUND the table, content and slightly too full, sipping our wine and waiting for the fondants au chocolat to come out of the oven. I'm sticking to water, though. I can't tolerate alcohol at all. I'm among the 30–40 percent of Southeast Asians who are genetically condemned to turn very red and get very drunk after even the tiniest amount of alcohol. That doesn't always stop them from drinking. A German friend of ours who lives in China tells us it's hard for him to keep up with the ungodly amounts of alcohol his Chinese colleagues drink during business dinners. Although many Chinese people can't hold their liquor any better than I can, they'll regularly drink until practically comatose, acting as though they have no other choice—despite being adults.

I admit that Matthias and I always stock up on alcohol when we invite people over. What kind of hosts would we be otherwise?

For chemists, "alcohol" is an entire class of substances. The alcohol *we're* talking about is a very specific alcohol, namely *ethanol*. All alcohols are poisonous, some more than others. But they aren't the only problem; the by-products in the metabolism are also an issue. Alcohols are usually broken down through oxidation, first to aldehyde, and then to carboxylic acid. *Methanol*, ethanol's brother with one fewer carbon atoms, would be less poisonous if our bodies didn't oxidize it to formaldehyde, which causes blindness. Drinking isopropyl, which is used in many disinfectants, can lead to breathing difficulties and circulatory collapse.

All alcohols are lethal at the right dosage. Ethanol is relatively tolerable, compared with other alcohols, but only to a point. Vomiting is the body's mechanism against ethanol poisoning. This can be dangerous too; if you're no longer in control of your faculties, you might choke on your own vomit.

Even if you don't drink yourself unconscious, you can still cause damage. Alcohol is metabolized in the liver. The liver's job is to get rid of toxic substances. Excessive alcohol consumption can overload and damage the liver—and that's just one of the incredibly long list of long-term health impacts. Your heart and digestive system will also be thankful if you refrain from drinking too much. And we haven't even touched on the poor decisions we make when inebriated.

So I consider my alcohol intolerance to be a blessing, and I hope to pass my mutated gene on to my children too.

202 | CHEMISTRY FOR BREAKFAST

That might sound mean, but believe me, you're not missing anything if you've never drunk alcohol. Many people pity me anyway.

Whether you pity me or not, my mutated gene is extremely fascinating. To understand it, we first need to know what exactly happens when ethanol enters the body.

Ethanol is oxidized to a substance called *acetaldehyde* with the aid of an enzyme called *alcohol dehydrogenase* (ADH for short). Acetaldehyde is at least as bad for your health as ethanol. It's a mutagen, a substance that damages the DNA and can cause cancer. Acetaldehyde is probably the reason why scientific studies continue to observe links between alcohol consumption and various forms of cancer. Accordingly, the body tries to get rid of this harmful substance as quickly as possible. The acetaldehyde is further oxidized in the metabolism to *acetic acid* or *acetate*, the salt of acetic acid. Only then is danger averted; the body has no problem excreting acetate or processing it further and converting it to energy. This is why alcohol has so many calories, by the way.

A second enzyme is also involved in oxidizing acetaldehyde to acetate. For most Europeans, this enzyme is called *aldehyde dehydrogenase* 2, or ALDH2. But for me and many other Southeast Asians (and a few people from other global regions), this enzyme looks different—and that's a huge problem.

To understand the problem, we first need to know what enzymes generally look like. I've discussed enzymes at various points throughout this book, but talking about alcohol is the perfect opportunity to give a few key details. Enzymes are proteins, so they are made of amino acids. Amino acids are small molecules built mainly on carbon, oxygen, hydrogen, and nitrogen. In total, our body makes proteins using twenty different amino acids:

Alanine   Glycine   Isoleucine   Leucine   Proline

Valine   Phenylalanine   Tryptophan   Tyrosine   Aspartic acid

Glutamate   Arginine   Histidine   Lysine   Serine

Threonine   Cysteine   Methionine   Asparagine   Glutamine

This diagram may seem a little overwhelming at first, but when you think that all our proteins are made from just these twenty building blocks—and that these proteins control pretty much all the biological processes in our bodies—it actually seems like an astonishingly small number.

In a protein, amino acids are linked to form a very long chain—for example, the ALDH2 enzyme comprises a chain of five hundred amino acids in a very specific sequence. This chain doesn't just lie around any old way; it's precisely folded like a piece of origami. Like water molecules, these chains can also form hydrogen bonds, one part of the chain linked to another, and the chain folds accordingly. At first glance, it all usually looks like a chaotic tangle. Each protein has a characteristic three-dimensional structure (depending on the type of amino acids and the order in which they occur), and this spatial structure determines its function.

This is a very complex matter, even for scientists who work with proteins every day. To ensure that everyone knows what they're talking about, they divide the structure into various aspects. The chemical composition of the chain is the *primary structure*; this focuses solely on the amino acid sequence, how the building blocks are arranged. The secondary, tertiary, and quaternary structures describe how the chain twists, folds, and congregates.

It's a bit like a Lego tower. If you focus on which bricks are used and in which order, that's the primary structure. The tower's three-dimensional form is its superstructure. Lego is too simple an analogy for proteins; if you swap a yellow Lego brick for a green one, the tower is still a tower. And you can swap two Lego bricks without the shape of the tower changing. Amino acids, however, are such sophisticated building blocks that even the tiniest variation in the primary sequence can have a huge impact on the superstructure.

My "broken" ALDH2 enzyme is a good example of this. There's something different about the amino acid at position 487 in the five-hundred-amino-acid chain—a single building block has been exchanged. And this one, tiny amino acid alters the hydrogen bonds within the enzyme, changing the superstructure. As a consequence, my enzyme can't break down acetaldehyde. It's inactive.

This doesn't mean that acetaldehyde isn't oxidized to acetic acid at all, just that the reaction is very, very slow. Acetaldehyde builds up in my body after just a few sips of wine. My body doesn't like this at all, responding with nausea, a rapid heartbeat, and a weird, lobster-red tinge to the skin, particularly in the face. The first time I tried alcohol in my youth, I thought I was allergic. The symptoms are termed the "Asian flush"—reason enough for me to avoid drinking completely.

I did get drunk once, and I don't intend to repeat the experience. I studied in Mainz, one of the heartlands of German Carnival. You might not be familiar with German Carnival— there are costumes and, first and foremost, a lot of drinking. (Yes, Oktoberfest is not the only alcohol-centered German festival.) It's unbearable if you're sober. I don't generally dislike being the only sober one in a crowd of drunks; I grew up in a typical small town with nothing much to do on the weekends apart from drink. But Carnival takes drunkenness to another level. What's interesting is that whenever someone asks me why I don't like Carnival, and I say that I can't drink alcohol, they always understand. "Yeah, I get it. It's awful without alcohol." I mean, what does it say about Carnival that nobody wants to do it sober?

Anyway, after fleeing Mainz for my first few Carnival seasons, I decided that I should stick around and go all out at least one time. After the morning parade and a few hours of clubbing in broad daylight, we ended up in a bar full of "old people" in the early evening. Looking back, they were probably only as old as I am now. I was twenty, and the "old people" were singing the popular Carnival song "Ich hab ne Zwiebel auf dem Kopf ich bin ein Döner," which translates to "I have an onion on my head, I'm a doner kebab"—see why this is for drunks only?? I couldn't leave yet; I'd paid ten euros to get in, a fortune for a student. I was trapped! So I decided the time had come to get drunk.

This proved more complicated than I expected. I tried various drinks but passed each one to my friends after just a sip; the alcohol tasted unbearably disgusting. Ethanol is an organic solvent, and that's exactly how it tastes to me. High-proof alcohol can downright burn because ethanol bonds to heat-sensitive receptors—the same ones activated by the capsaicin molecules in chilies. These receptors pass on the feeling of heat to the brain, where it's translated into sensations of pain. While capsaicin imitates direct heat, ethanol simply makes the heat receptors more sensitive. It lowers the threshold temperature, the heat receptors suddenly find your own body temperature too hot, and you feel a burning on your tongue. Let's get back to my attempt to get drunk.

Half an hour and a few sips later—I hadn't even managed half a drink overall—I desperately needed some fresh air. I'd barely taken a breath before I surprised myself by puking on the doorstep. I felt pretty good after that and told my friends that the evening was only just beginning! Less than ten minutes later, all I wanted was to go home. Snickering, my friends took me back—I could barely walk unaided—and I threw up again at home. I was in bed by 7 p.m. That's what happens when one of your alcohol metabolism enzymes doesn't work correctly.

ALCOHOL IS EVEN poisonous for the yeasts that produce it. During the fermentation process, the yeasts feed on sugars and carbohydrates, and one of the things they produce is ethanol. By the time the alcohol content gets to 15 percent, things have become pretty unpleasant for the yeast cells. Tragically, they die at the hands of their own metabolic product. This is why high-proof alcohol can only be made through distillation, which uses evaporation and condensation to concentrate the alcohol content.

I MAY NOT get to experience the rush of alcohol, but being around drunk people when you're sober is quite the spectacle. Their lack of control is pretty disconcerting, but over the millennia we've become accustomed to alcohol and everything the drug entails. Interesting when you consider how much we demonize all other forms of chemically induced intoxication. On the other hand, I have to admit that an evening with wine is much funnier than an evening without.

Dino in particular seems to benefit from the alcohol. Over the last couple of hours, he's proven to be extremely witty, something he hides behind his silence when sober. Why is that? What exactly does this ethanol molecule do to our bodies to make us more confident and less inhibited? Let's take a look at the chemistry of alcohol intoxication.

The ethanol is absorbed in the stomach and small intestine and enters the bloodstream. Most of it is sent to the liver, which begins enzymatic breakdown. A small amount escapes via the lungs—this is what makes your breath smell of alcohol, together with acetaldehyde. Not nice for the people around you, but very practical for law enforcement. These processes are simply the body's sensible attempt to get rid of the alcohol as quickly as possible. However, it can't usually keep up with our average drinking speed, and the excess ethanol gets into the brain via the bloodstream. And then the fun really begins.

Alcohol affects the brain in a similar way to sedatives or an anesthetic—a strange thought when you imagine raucous drinkers dancing on tables. Although drunk people lose their inhibitions, alcohol actually inhibits the function of neurons—more precisely, it inhibits communication between our nerve cells, which communicate via neurotransmitters (remember the serotonin in Chapter 7?). I'm going to introduce you to two more neurotransmitters to help you better understand your drunken self in the future.

The last diagram included a molecule called *glutamate* (or *glutamic acid*), one of the twenty amino acids that form proteins. Glutamate also works as an *excitatory neurotransmitter*. When glutamate bonds to its receptor, it activates communication between the nerve cells; more signals are sent.

Its counterpart is the inhibitory neurotransmitter GABA (gamma-aminobutyric acid)—the neurotransmitter whose release is stimulated by oxytocin, the (possible) cuddle hormone. When GABA bonds to its receptor, it inhibits communication and ensures that fewer signals are transmitted.

Glutamate/glutamic acid
(= excitatory)

GABA
(= inhibitory)

We've spoken a lot about receptors in the brain (remember my parking space comparison?). Now let's take a closer look, at least in the context of drinking. Like so many things in our bodies, receptors are made up of proteins. Visualize them as tunnels or channels that are normally closed. When the right neurotransmitter bonds to the receptor, these channels open for a moment and ions (such as sodium, potassium, calcium, or chloride ions) can pass through. Just like in a cell or mobile phone battery, when these charged particles pass through, a voltage is created that enables the nerve cells to send electrical signals. If these ions have a positive charge (cations), the nerve cell fires signals; this is what happens with the glutamate receptor. If the ions have a negative charge (anions), the signals are suppressed; this is what happens with the GABA receptor.

Then ethanol enters the scene and messes everything up. The ethanol molecule interacts with both the glutamate receptor and the GABA receptor. The flow of ions to the glutamate receptor is inhibited. Glutamate's excitatory effect is reduced by alcohol, meaning that the nerve cells send fewer signals. Meanwhile, ethanol ensures that the channels at the GABA receptor remain open for longer, meaning that more ions flow through. GABA's inhibitory effect is increased by alcohol, which also means that the nerve cells send fewer signals.

So alcohol slows our brains down in two different ways. This would explain why it prompts us to shout and dance on tables; among other things, the reduced brain activity inhibits social anxiety and general self-control. It also affects our motor skills. If our neurons are communicating less, then simple tasks like walking in a straight line are harder than usual. We also start to slur our words and react slowly. And we don't necessarily make the smartest decisions of our lives—if you've ever been drunk, I'm sure you have your own anecdotes to confirm this. As a general rule, we think less, are less observant, and remember less.

Under normal circumstances, GABA's inhibiting effect makes it a very important neurotransmitter. Obviously we need active brain cells, but less is more. GABA helps us to organize information and differentiate between stimuli. Without this inhibiting effect, we'd struggle to think clearly and fall to pieces from sensory overload. This is why medications that increase GABA concentration are prescribed to people with epilepsy. I sometimes wonder whether their reduced brain activity helps drunks to think more clearly (if very little). I can't relate to it myself, but it does fit with the tendency of drunk people to repeat the same idea or realization over and over again.

Alcohol influences the brain in many other ways too. Ethanol increases the release of *dopamine*, a wonderful yet insidious neurotransmitter. This molecule is involved in so many things, including movement, learning, attention, and emotions. It's one of the core neurotransmitters in our reward system: whenever we do something that brings us joy, dopamine is released—and we want more. Too much dopamine leads to impulsive and addictive behavior, and it even plays a significant role in schizophrenia. Once dopamine is released, self-control is required to resist the rush of happiness and the urge to keep on refilling your wineglass—tricky if the relevant part of the brain has already been incapacitated by the alcohol. And so it's not until the next morning that you regret how much you drank.

MATTHIAS, CHRISTINE, AND Torben are only a little tipsy—after all, we've eaten well. Ethanol gets into the bloodstream through the stomach and small intestine, so a full belly can slow its intake.

That is, of course, unless you suffer from *auto-brewery syndrome*, an incredible and rare condition with only a few medical case studies.

The story begins in 2004 with a middle-aged American man who, after a foot operation and treatment with anti-biotics, experienced a sudden and significant drop in his alcohol tolerance. He'd get totally hammered after two small beers, and sometimes he felt drunk even without alcohol. His wife, a nurse, started to document his alcohol levels using a Breathalyzer; blood alcohol concentrations of 0.3 percent weren't unusual. In Germany, drivers with a blood alcohol concentration of 0.11 percent or above are considered absolutely unfit to drive; drivers with a concentration of 0.03 percent are considered relatively unfit to drive. In the United States, the limit for drivers is 0.08 percent. This guy was in serious trouble. The couple couldn't account for his alcohol levels and started looking for hidden alcohol, for example in chocolates, but couldn't find any explanation. They must have really trusted each other, given that she didn't accuse him of drinking in secret. In 2009, the man was taken to the emergency room with a life-threatening concentration of 0.37 percent, despite not having drunk any alcohol (or so he claimed). The physicians didn't believe a word he said and assumed that he must be a closet alcoholic.

One year later, the man returned to the hospital for a colonoscopy and his physicians discovered something very unusual: a yeast called *Saccharomyces cerevisiae*, also known as baker's yeast—or brewer's yeast. As the name suggests, this yeast is used to brew beer, but it has no place in our digestive system. Did the man have a brewery in his stomach? It gives a whole new meaning to "beer belly."

The physicians started to investigate. In April 2010, the man was admitted to hospital for twenty-four hours so his stomach could be examined. Naturally, his belongings were searched thoroughly to make sure he hadn't smuggled in any alcohol. Skepticism remained. The physicians gave him water with glucose and various carbohydrate snacks. By the afternoon, the poor man was actually drunk—with a blood alcohol concentration of 0.12 percent. His brewer's yeast had converted the carbohydrates into ethanol. Appropriately, this phenomenon was named auto-brewery syndrome. It's so rare that, apart from individual case studies, there are no detailed scientific studies. As far as I know, the man is feeling better; the problem is under control thanks to a combination of fungicide treatment and a low-carb diet. I'm so glad I don't have the condition—can you imagine?

"MAI," DINO SAYS, "I think what you do is remarkable, getting people excited about chemistry... but I do have one criticism."

Christine leans forward in curiosity. First jokes, now criticism. Alcohol really can work miracles.

"You can't just blindly motivate young people," Torben says. "They can't all study chemistry! That would be a disaster!"

We laugh, but Torben is actually serious—and honestly, this is a serious issue. Let me explain.

Getting people excited about chemistry is one of my goals, and my passion. I receive lots of wonderful messages and comments from young people, telling me that they'd never had any interest in science before, let alone chemistry, but have acquired a taste for it after watching my videos. Some write that they're about to start vocational training or university studies and generously name me as their inspiration or motivation. In turn, messages like these are my greatest motivation.

Why exactly is this important?

Again and again, we hear it said that we need more STEM. STEM stands for *science, technology, engineering, and mathematics* (so everything that I find fun). And there aren't enough STEM specialists! Apparently. A skills shortage isn't a particularly compelling reason. Plus, the situation on the labor market may well change. For years, over ten thousand people in Germany have chosen to study chemistry each year. There were over eleven thousand new students in 2017, more than ever. Over two thousand chemists received their doctorates in 2017. There's still a desperate need to fill jobs such as lab technicians in the chemistry sector, but there's no shortage of chemistry PhDs at the moment.

"I don't necessarily want to encourage people to train in chemistry or study it," I say. "That's just a side effect of my actual mission."

"MAI MISSION!!" Matthias and Christine shout in unison, raising their glasses.

They rag on me sometimes for being "on a mission." That's a bit too melodramatic and serious for my taste. But I am serious, and they both know that and support me.

In my opinion, it's far too shortsighted to say that we need more STEM due to a lack of specialists. That would mean that young people only need to be interested in STEM until all vacancies are filled. By that way of thinking, nobody needs to be interested in chemistry at the moment because the labor market has plenty of chemists with doctorates. Obviously, that's hooey. I say we need more STEM because science, technology, engineering, and mathematics are crucial to our lives and we need to understand them! That doesn't mean you have to study them.

It doesn't matter if you dropped chemistry in high school (I forgive you). It doesn't matter if you think physics is cooler than chemistry (I forgive you too, but remember that only beginners draw boundaries between sciences). And if you'd rather be a carpenter or study art history, great! Chemistry can just as easily be a hobby as soccer or guitar can. Everyone should know more about chemistry!

However, I don't want people to learn about chemistry simply so that they know more. You've learned a lot throughout this book; you've learned about the particle model and thermodynamics, the shell model and the octet rule, chemical bonds and hydrogen bonds, oxidation and reduction, neurotransmitters and hormones, surfactants and fluorides, theobromine and caffeine. I could tell the story of my day all over again, but with completely different chemical examples. You could do the same with biology or physics. It doesn't actually matter what information you take away from this book; what matters is that you're infected with the scientific spirit. And that, essentially, is my mission—more scientific spirit! Chemistry is my weapon of choice, but there are plenty of others. All sciences are united by the same spirit. I'd better explain what I mean:

Scientific spirit means not taking anything for granted and observing the world through fresh eyes. Looking for the miraculous in the familiar. Scientific spirit is the moment that you pick up your daily cup of coffee and think, "Hmm. All molecules. Amazing."

Scientific spirit means recognizing the beauty of inner workings; seeing flowers through Richard Feynman's eyes; realizing that every new scientific discovery brings more questions, more wonders, and more beauty.

Scientific spirit means celebrating researchers who conduct double-blind studies because they know there's always a chance that personal expectations can cloud critical thinking.

Scientific spirit is insatiable curiosity that can't be curbed by even the world's stinkiest molecule.

Scientific spirit means delighting in complexity and resisting simple answers. Anyone who discovers chemistry and enjoys understanding chemical connections not only will enhance their life and day-to-day existence, but will inevitably come to delight in complexity.

Scientific spirit is a love of facts and figures. This includes being aware of the biases we all have, scrutinizing our personal opinions, and being prepared to change them if the facts demand it. Facts and personal opinions mustn't be treated as equal. Some political debates leave me with my head in my hands, asking why it's so difficult to be both emotional and factual. And when I watch some of my fellow scientists, I ask why it's so difficult to be both factual and passionate. Facts and emotions aren't mutually exclusive. What I want is a passion for facts!

"I'LL DRINK TO that," Christine says, raising her glass. "To more passion for facts!"

"To more passion for facts!" we echo. Our wineglasses clink, and the atoms collide and begin to vibrate. The sound waves spread, sending the air swirling through the room, and the molecules dance around our heads.

Bibliography

HERE YOU WILL find the most important sources, particularly for content that can't be found in standard textbooks. Sources are listed in the order they appear in the text.

Chapter 1

Lewy, A. J., T. A. Wehr, F. K. Goodwin, D. A. Newsome, and S. P. Markey. "Light suppresses melatonin secretion in humans." *Science* 210, no. 4475 (1980): 1267–1269. https://doi.org/10.1126/science.7434030.

Herman, J. P., J. M. McKlveen, S. Ghosal, B. Kopp, A. Wulsin, R. Makinson, J. Scheimann, and B. Myers. "Regulation of the hypothalamic-pituitary-adrenocortical stress response." *Comprehensive Physiology* 6, no. 2 (2016): 603. https://doi.org/10.1002/cphy.c150015.

McEwen, B. S., and E. Stellar. "Stress and the individual: mechanisms leading to disease." *Archives of Internal Medicine* 153, no. 18 (1993): 2093–2101. https://doi.org/10.1001/archinte.1993.00410180039004.

Wilhelm, I., J. Born, B. M. Kudielka, W. Schlotz, and S. Wüst. "Is the cortisol awakening rise a response to awakening?" *Psychoneuroendocrinology* 32, no. 4 (2007): 358–366. https://doi.org/10.1016/j.psyneuen.2007.01.008.

Wüst, S., J. Wolf, D. H. Hellhammer, I. Federenko, N. Schommer, and C. Kirschbaum. "The cortisol awakening response—normal values and confounds." *Noise & Health* 2, no. 7 (2000): 79–88. http://www.noiseandhealth.org/text.asp?2000/2/7/79/31739.

Wren, M. A., R. T. Dauchy, J. P. Hanifin, M. R. Jablonski, B. Warfield, G. C. Brainard, D. E. Blask, S. M. Hill, T. G. Ooms, and R. P. Bohm Jr. "Effect of different spectral transmittances through tinted animal cages on circadian metabolism and physiology in Sprague-Dawley rats." *Journal of the American Association for Laboratory Animal Science* 53, no. 1 (2014): 44–51. https://www.ncbi.nlm.nih.gov/pmc/articles/PMC3894647/.

van Geijlswijk, I. M., H. P. Korzilius, and M. G. Smits. "The use of exogenous melatonin in delayed sleep phase disorder: a meta-analysis." *Sleep* 33, no. 12 (2010): 1605–1614. https://doi.org/10.1093/sleep/33.12.1605.

Claustrat, B., and J. Leston. "Melatonin: physiological effects in humans." *Neurochirurgie* 61, no. 2–3 (2015): 77–84. https://doi.org/10.1016/j.neuchi .2015.03.002.

Zisapel, N. "New perspectives on the role of melatonin in human sleep, circadian rhythms and their regulation." *British Journal of Pharmacology* 175, no. 16 (2018): 3190–3199. https://doi.org/10.1111/bph.14116.

Lovallo, W. R., T. L. Whitsett, M. al'Absi, B. H. Sung, A. S. Vincent, and M. F. Wilson. "Caffeine stimulation of cortisol secretion across the waking hours in relation to caffeine intake levels." *Psychosomatic Medicine* 67, no. 5 (2005): 734–739. https://doi.org/10.1097/01.psy.0000181270.20036.06.

Huang, R. C. "The discoveries of molecular mechanisms for the circadian rhythm: the 2017 Nobel Prize in Physiology or Medicine." *Biomedical Journal* 41, no. 1 (2018): 5–8. https://doi.org/10.1016/j.bj.2018.02.003.

Stothard, E. R., A. W. McHill, C. M. Depner, B. R. Birks, T. M. Moehlman, H. K. Ritchie, J. R. Guzzetti, et al. "Circadian entrainment to the natural light-dark cycle across seasons and the weekend." *Current Biology* 27, no. 4 (2017): 508–513. https://doi.org/10.1016/j.cub.2016.12.041.

Chapter 2

American Cancer Society. "Perfluorooctanoic acid (PFOA), Teflon, and related chemicals." March 4, 2020, https://www.cancer.org/cancer/cancer-causes /teflon-and-perfluorooctanoic-acid-pfoa.html.

Chapter 3

Meyer-Lückel, Henrik, Sebastian Paris, and Kim Ekstrand, eds. *Karies: Wissenschaft und Klinische Praxis*. Stuttgart, Germany: Georg Thieme Verlag, 2012.

Choi, A. L., S. Guifan, Y. Zhang, and P. Grandjean. "Developmental fluoride neurotoxicity: a systematic review and meta-analysis." *Environmental Health Perspectives* 120, no. 10 (2012): 1362–1368. https://doi.org/10.1289/ehp.1104912.

Bashash, M., D. Thomas, H. Hu, E. Angeles Martinez-Mier, B. N. Sanchez, N. Basu, K. E. Peterson, et al. "Prenatal fluoride exposure and cognitive outcomes in children at 4 and 6–12 years of age in Mexico." *Environmental Health Perspectives* 125, no. 9 (2017): 097017. https://doi.org/10.1289/EHP655.

EFSA Panel on Dietetic Products, Nutrition, and Allergies (NDA). "Scientific opinion on dietary reference values for fluoride." *EFSA Journal* 11, no. 8 (2013): 3332. https://doi.org/10.2903/j.efsa.2013.3332.

Chapter 4

Tremblay, M. S., R. C. Colley, T. J. Saunders, G. N. Healy, and N. Owen. "Physiological and health implications of a sedentary lifestyle." *Applied Physiology, Nutrition, and Metabolism* 35, no. 6 (2010): 725–740. https://doi.org/10.1139/H10-079.

Baddeley, B., S. Sornalingam, and M. Cooper. "Sitting is the new smoking: where do we stand?" *British Journal of General Practice* 66, no. 646 (2016): 258. https://doi.org/10.3399/bjgp16X685009.

World Health Organization (WHO). *Noncommunicable Diseases: Progress Monitor 2017*. Geneva: WHO, 2017.

Forouzanfar, M. H., A. Afshin, L. T. Alexander, H. R. Anderson, Z. A. Bhutta, S. Biryukov, M. Brauer, et al. "Global, regional, and national comparative risk assessment of 79 behavioural, environmental and occupational, and metabolic risks or clusters of risks, 1990–2015: a systematic analysis for the Global Burden of Disease Study 2015." *The Lancet* 388, no. 10053 (2016): 1659–1724. https://doi.org/10.1016/S0140-6736(16)31679-8.

Chau, J. Y., C. Bonfiglioli, A. Zhong, Z. Pedisic, M. Daley, B. McGill, and A. Bauman. "Sitting ducks face chronic disease: an analysis of newspaper coverage of sedentary behaviour as a health issue in Australia 2000–2012." *Health Promotion Journal of Australia* 28, no. 2 (2017): 139–143. https://doi.org/10.1071/HE16054.

Ekelund, U., J. Steene-Johannessen, W. J. Brown, M. W. Fagerland, N. Owen, K. E. Powell, A. Bauman, I. M. Lee, and the Lancet Sedentary Behaviour Working Group. "Does physical activity attenuate, or even eliminate, the detrimental association of sitting time with mortality? A harmonised meta-analysis of data from more than 1 million men and women." *The Lancet* 388, no. 10051 (2016): 1302–1310. https://doi.org/10.1016 /s0140-6736(16)30370-1.

O'Donovan, G., I. M. Lee, M. Hamer, and E. Stamatakis. "Association of 'weekend warrior' and other leisure time physical activity patterns with risks for all-cause, cardiovascular disease, and cancer mortality." JAMA *Internal Medicine* 177, no. 3 (2017): 335–342. https://doi.org/10.1001 /jamainternmed.2016.8014.

Martin, A., C. Fitzsimons, R. Jepson, D. H. Saunders, H. P. van der Ploeg, P. J. Teixeira, C. M. Gray, and N. Mutrie. "Interventions with potential to reduce sedentary time in adults: systematic review and meta-analysis." *British Journal of Sports Medicine* 49, no. 16 (2015): 1056–1063. http://dx.doi.org/10.1136 /bjsports-2014-094524.

Stamatakis, E., R. M. Pulsford, E. J. Brunner, A. R. Britton, A. E. Bauman, S. J. Biddle, and M. Hillsdon. "Sitting behaviour is not associated with incident diabetes over 13 years: the Whitehall II cohort study." *British Journal of Sports Medicine* 51, no. 10 (2017): 818–823. http://dx.doi.org/10.1136 /bjsports-2016-096723.

Marmot, M., and E. Brunner. "Cohort profile: the Whitehall II study." *International Journal of Epidemiology* 34, no. 2 (2005): 251–256. https://doi.org /10.1093/ije/dyh372.

Biswas, A., P. I. Oh, G. E. Faulkner, R. R. Bajaj, M. A. Silver, M. S. Mitchell, and D. A. Alter. "Sedentary time and its association with risk for disease incidence, mortality, and hospitalization in adults: a systematic review and meta-analysis." *Annals of Internal Medicine* 162, no. 2 (2015): 123–132. https:// doi.org/10.7326/M14-1651.

van Uffelen, J. G., J. Wong, J. Y. Chau, H. P. van der Ploeg, I. Riphagen, N. D. Gilson, N. W. Burton, et al. "Occupational sitting and health risks: a systematic review." *American Journal of Preventive Medicine* 39, no. 4 (2010): 379–388. https://doi.org/10.1016/j.amepre.2010.05.024.

Stamatakis, E., N. Coombs, A. Rowlands, N. Shelton, and M. Hillsdon. "Objectively-assessed and self-reported sedentary time in relation to multiple socioeconomic status indicators among adults in England: a cross-sectional study." BMJ Open 4, no. 11 (2014): e006034. http://dx.doi.org /10.1136/bmjopen-2014-006034.

Grøntved, A., and F. B. Hu. "Television viewing and risk of type 2 diabetes, cardiovascular disease, and all-cause mortality: a meta-analysis." JAMA 305, no. 23 (2011): 2448–2455. https://doi.org/10.1001/jama.2011.812.

Stamatakis, E., M. Hillsdon, G. Mishra, M. Hamer, and M. Marmot. "Television viewing and other screen-based entertainment in relation to multiple socioeconomic status indicators and area deprivation: the Scottish Health Survey 2003." Journal of Epidemiology and Community Health 63, no. 9 (2009): 734–740. http://dx.doi.org/10.1136/jech.2008.085902.

Hamer, M., E. Stamatakis, and G. D. Mishra. "Television- and screen-based activity and mental well-being in adults." American Journal of Preventive Medicine 38, no. 4 (2010): 375–380. https://doi.org/10.1016/j.amepre.2009.12.030.

Pearson, N., and S. J. Biddle. "Sedentary behavior and dietary intake in children, adolescents, and adults: a systematic review." American Journal of Preventive Medicine 41, no. 2 (2011): 178–188. https://doi.org/10.1016 /j.amepre.2011.05.002.

Scully, M., H. Dixon, and M. Wakefield. "Association between commercial television exposure and fast-food consumption among adults." Public Health Nutrition 12, no. 1 (2009): 105–110. https://doi.org/10.1017/S1368980008002012.

Chapter 5

Liljenquist, K., C. B. Zhong, and A. D. Galinsky. "The smell of virtue: clean scents promote reciprocity and charity." Psychological Science 21, no. 3 (2010): 381–383. https://doi.org/10.1177/0956797610361426.

Vohs, K. D., J. P. Redden, and R. Rahinel. "Physical order produces healthy choices, generosity, and conventionality, whereas disorder produces creativity." Psychological Science 24, no. 9 (2013): 1860–1867. https://doi.org /10.1177/0956797613480186.

Open Science Collaboration. "Estimating the reproducibility of psychological science." Science 349, no. 6251 (2015): aac4716. https://doi.org/10.1126/science.aac4716.

Price, D. D., D. G. Finniss, and F. Benedetti. "A comprehensive review of the placebo effect: recent advances and current thought." *Annual Review of Psychology* 59 (2008): 565–590. https://doi.org/10.1146/annurev.psych.59.113006.095941.

Jewett, D. L., G. Fein, and M. H. Greenberg. "A double-blind study of symptom provocation to determine food sensitivity." *New England Journal of Medicine* 323, no. 7 (1990): 429–433. https://doi.org/10.1056/NEJM199008163230701.

Benedetti, F., M. Lanotte, L. Lopiano, and L. Colloca. "When words are painful: unraveling the mechanisms of the nocebo effect." *Neuroscience* 147, no. 2 (2007): 260–271. https://doi.org/10.1016/j.neuroscience.2007.02.020.

Chapter 6

Yogeshwar, R. "Und was bringt's mir?" (translation). *Spektrum.de*, May 7, 2018, https://www.spektrum.de/kolumne/und-was-bringts-mir/1563312.

Rohrig, B. "Smartphones: smart chemistry." *ChemMatters*, April/May 2015, https://www.acs.org/content/acs/en/education/resources/highschool/chemmatters/past-issues/archive-2014-2015/smartphones.html.

Buchmann, I. *Batteries in a Portable World: A Handbook on Rechargeable Batteries for Non-engineers.* Richmond, Canada: Cadex Electronics Inc., 2001.

Braga, M. H., C. M. Subramaniyam, A. J. Murchison, and J. B. Goodenough. "Nontraditional, safe, high voltage rechargeable cells of long cycle life." *Journal of the American Chemical Society* 140, no. 20 (2018): 6343–6352. https://doi.org/10.1021/jacs.8b02322.

Chapter 7

Asberg, M., P. Thorén, L. Träskman, L. Bertilsson, and V. Ringberger. "'Serotonin depression'—a biochemical subgroup within the affective disorders?" *Science* 191, no. 4226 (1976): 478–480. https://doi.org/10.1126/science.1246632.

Song, F., N. Freemantle, T. A. Sheldon, A. House, P. Watson, A. Long, and J. Mason. "Selective serotonin reuptake inhibitors: meta-analysis of efficacy and acceptability." *British Medical Journal* 306, no. 6879 (1993): 683–687. https://doi.org/10.1136/bmj.306.6879.683.

Owens, M. J., and C. B. Nemeroff. "Role of serotonin in the pathophysiology of depression: focus on the serotonin transporter." *Clinical Chemistry* 40, no. 2 (1994): 288–295. http://citeseerx.ist.psu.edu/viewdoc/download?doi=10.1.1.539.5231&rep=rep1&type=pdf.

Whittington, C. J., T. Kendall, P. Fonagy, D. Cottrell, A. Cotgrove, and E. Boddington. "Selective serotonin reuptake inhibitors in childhood depression: systematic review of published versus unpublished data." *The Lancet* 363, no. 9418 (2004): 1341–1345. https://doi.org/10.1016 /s0140-6736(04)16043-1.

Fergusson, D., S. Doucette, K. C. Glass, S. Shapiro, D. Healy, P. Hebert, and B. Hutton. "Association between suicide attempts and selective serotonin reuptake inhibitors: systematic review of randomised controlled trials." *British Medical Journal* 330, no. 7488 (2005): 396. https://doi.org/10.1136/bmj.330.7488.396.

Risch, N., R. Herrell, T. Lehner, K. Y. Liang, L. Eaves, J. Hoh, A. Griem, M. Kovacs, J. Ott, and K. R. Merikangas. "Interaction between the serotonin transporter gene (5-HTTLPR), stressful life events, and risk of depression: a meta-analysis." *JAMA* 301, no. 23 (2009): 2462–2471. https://doi.org/10.1001 /jama.2009.878.

Karg, K., M. Burmeister, K. Shedden, and S. Sen. "The serotonin transporter promoter variant (5-HTTLPR), stress, and depression meta-analysis revisited: evidence of genetic moderation." *Archives of General Psychiatry* 68, no. 5 (2011): 444–454. https://doi.org/10.1001/archgenpsychiatry.2010.189.

Aguilar, F., H. Autrup, S. Barlow, L. Castle, R. Crebelli, W. Dekant, K. H. Engel, et al. "Assessment of the results of the study by McCann et al. (2007) on the effect of some colours and sodium benzoate on children's behaviour." *The* EFSA *Journal* 660 (2008): 1–54. https://doi.org/10.2903/j.efsa.2008.660.

McCann, D., A. Barrett, A. Cooper, D. Crumpler, L. Dalen, K. Grimshaw, E. Kitchin, et al. "Food additives and hyperactive behaviour in 3-year-old and 8/9-year-old children in the community: a randomised, double-blinded, placebo-controlled trial." *The Lancet* 370, no. 9598 (2007): 1560–1567. https://doi.org/10.1016/s0140-6736(07)61306-3.

Schab, D. W., and N. H. Trinh. "Do artificial food colors promote hyperactivity in children with hyperactive syndromes? A meta-analysis of double-blind placebo-controlled trials." *Journal of Developmental & Behavioral Pediatrics* 25, no. 6 (2004): 423–434. https://doi.org/10.1097/00004703-200412000-00007.

Watson, R. "European agency rejects links between hyperactivity and food additives." *British Medical Journal* 336, 7646 (2008): 687. https://doi.org /10.1136/bmj.39527.401644.DB.

EFSA Panel on Food Additives and Nutrient Sources (ANS). "Scientific Opinion on the re-evaluation of benzoic acid (E 210), sodium benzoate (E 211), potassium benzoate (E 212) and calcium benzoate (E 213) as food additives." EFSA *Journal* 14, no. 4 (2016): 4433. https://doi.org/10.2903/j.efsa.2016.4433.

Chapter 8

Letter from Erick M. Carreira: The original source cannot be traced. The letter has been available online since it was leaked. A photograph of the letter can be found at: http://www.chemistry-blog.com/tag/carreira-letter/.

Chapter 9

Zeng, X.-N., J. J. Leyden, H. J. Lawley, K. Sawano, I. Nohara, and G. Preti. "Analysis of characteristic odors from human male axillae." *Journal of Chemical Ecology* 17, no. 7 (1991): 1469–1492. https://doi.org/10.1007/BF00983777.

Fredrich, E., H. Barzantny, I. Brune, and A. Tauch. "Daily battle against body odor: towards the activity of the axillary microbiota." *Trends in Microbiology* 21, no. 6 (2013): 305–312. https://doi.org/10.1016/j.tim.2013.03.002.

Compound Interest. "The chemistry of body odours—sweat, halitosis, flatulence & cheesy feet." April 7, 2014, https://www.compoundchem.com/2014/04/07/the-chemistry-of-body-odours-sweat-halitosis-flatulence-cheesy-feet/.

Suarez, F. L., J. Springfield, and M. D. Levitt. "Identification of gases responsible for the odour of human flatus and evaluation of a device purported to reduce this odour." *Gut* 43, no. 1 (1998): 100–104. https://doi.org/10.1136/gut.43.1.100.

Fromm, E., and E. Baumann. "Ueber Thioderivate der Ketone." *Berichte der deutschen chemischen Gesellschaft* 22, no. 1 (1889): 1035–1045. https://doi.org/10.1002/cber.188902201224.

Baumann, E., and E. Fromm. "Ueber Thioderivate der Ketone." *Berichte der deutschen chemischen Gesellschaft* 22, no. 2 (1889): 2592–2599. https://doi.org/10.1002/cber.188902202151.

Krewski, D., R. A. Yokel, E. Nieboer, D. Borchelt, J. Cohen, J. Harry, S. Kacew, J. Lindsay, A. M. Mahfouz, and V. Rondeau. "Human health risk assessment for aluminium, aluminium oxide, and aluminium hydroxide." *Journal of Toxicology and Environmental Health, Part B* 10, S1 (2007): 1–269. https://doi.org/10.1080/10937400701597766.

Bundesinstitut für Risikobewertung. "Aluminiumhaltige Antitranspirantien tragen zur Aufnahme von Aluminium bei. Stellungnahme Nr. 007/2014." (2014). http://www.bfr.bund.de/cm/343/aluminiumhaltige -antitranspirantien-tragen-zur-aufnahme-von-aluminium-bei.pdf.

Scientific Committee on Consumer Safety (SCCS). "Opinion on the safety of aluminium in cosmetic products, preliminary version of 30-31 October 2019, final version of 03-04 March 2020, SCCS/1613/19." (2020). https://ec.europa.eu/health/sites/health/files/scientific_committees /consumer_safety/docs/sccs_o_235.pdf.

Bundesinstitut für Risikobewertung. "Neue Studien zu aluminiumhaltigen Antitranspirantien: Gesundheitliche Beeinträchtigungen durch Aluminium- Aufnahme über die Haut sind unwahrscheinlich. Stellungnahme 030/2020 des BfR vom 20." (2020). https://www.bfr.bund.de/cm/343/neue -studien-zu-aluminiumhaltigen-antitranspirantien-gesundheitliche -beeintr%c3%a4chtigungen-durch-aluminium-aufnahme-ueber-die -haut-sind-unwahrscheinlich.pdf.

Callewaert, C., E. De Maeseneire, F. M. Kerckhof, A. Verliefde, T. Van de Wiele, and N. Boon. "Microbial odor profile of polyester and cotton clothes after a fitness session." *Applied and Environmental Microbiology* 80, no. 21 (2014): 6611–6619. https://doi.org/10.1128/AEM.01422-14.

Chapter 10

Hampson, N. B., N. W. Pollock, and C. A. Piantadosi. "Oxygenated water and athletic performance." *JAMA* 290, no. 18 (2003): 2408–2409. https://doi.org/10.1001/jama.290.18.2408-a.

Eweis, D. S., F. Abed, and J. Stiban. "Carbon dioxide in carbonated beverages induces ghrelin release and increased food consumption in male rats: implications on the onset of obesity." *Obesity Research & Clinical Practice* 11, no. 5 (2017): 534–543. https://doi.org/10.1016/j.orcp.2017.02.001.

Vartanian, L. R., M. B. Schwartz, and K. D. Brownell. "Effects of soft drink consumption on nutrition and health: a systematic review and meta- analysis." *American Journal of Public Health* 97, no. 4 (2007): 667–675. https://doi.org/10.2105/AJPH.2005.083782.

Mourao, D. M., J. Bressan, W. W. Campbell, and R. D. Mattes. "Effects of food form on appetite and energy intake in lean and obese young adults."

International Journal of Obesity 31, no. 11 (2007): 1688–1695. https://doi.org/10.1038/sj.ijo.0803667.

Chapter 11

Baggott, M. J., E. Childs, A. B. Hart, E. De Bruin, A. A. Palmer, J. E. Wilkinson, and H. De Wit. "Psychopharmacology of theobromine in healthy volunteers." *Psychopharmacology* 228, no. 1 (2013): 109–118. https://doi.org/10.1007/s00213-013-3021-0.

Judelson, D. A., A. G. Preston, D. L. Miller, C. X. Muñoz, M. D. Kellogg, and H. R. Lieberman. "Effects of theobromine and caffeine on mood and vigilance." *Journal of clinical psychopharmacology* 33, no. 4 (2013): 499–506. https://doi.org/10.1097/JCP.0b013e3182905d24.

Mumford, G. K., S. M. Evans, B. J. Kaminski, K. L. Preston, C. A. Sannerud, K. Silverman, and R. R. Griffiths. "Discriminative stimulus and subjective effects of theobromine and caffeine in humans." *Psychopharmacology* 115, no. 1–2 (1994): 1–8. https://doi.org/10.1007/bf02244744.

Li, X., W. Li, H. Wang, D. L. Bayley, J. Cao, D. R. Reed, A. A. Bachmanov, et al. "Cats lack a sweet taste receptor." *The Journal of Nutrition* 136, no. 7 (2006): 1932S–1934S. https://doi.org/10.1093/jn/136.7.1932S.

Li, X., D. Glaser, W. Li, W. E. Johnson, S. J. O'Brien, G. K. Beauchamp, and J. G. Brand. "Analyses of sweet receptor gene (Tas1r2) and preference for sweet stimuli in species of Carnivora." *Journal of Heredity* 100, s1 (2009): 90–100. https://doi.org/10.1093/jhered/esp015.

Huth, P. J. "Do ruminant *trans* fatty acids impact coronary heart disease risk?" *Lipid Technology* 19, no. 3 (2007): 59–62. https://doi.org/10.1002/lite.200600021.

Food and Agriculture Organization (FAO). *Fats and Fatty Acids in Human Nutrition.* Food and Nutrition Paper 91. Rome: FAO, 2010. http://www.fao.org/3/a-i1953e.pdf.

Nishida, C., and R. Uauy. "WHO Scientific Update on health consequences of trans fatty acids: introduction." *European Journal of Clinical Nutrition* 63, s2 (2009): 1–4. https://doi.org/10.1038/ejcn.2009.13.

Simopoulos, A. P., A. Leaf, and N. Salem Jr. "Essentiality of and recommended dietary intakes for omega-6 and omega-3 fatty acids." *Annals of Nutrition and Metabolism* 43, no. 2 (1999): 127–130. https://doi.org/10.1159/000012777.

Servick, K. "The war on gluten." *Science* 360, no. 6391 (2018): 848–851. https://doi.org/10.1126/science.360.6391.848.

Catassi, C., J. C. Bai, B. Bonaz, G. Bouma, A. Calabrò, A. Carroccio, G. Castillejo, et al. "Non-celiac gluten sensitivity: the new frontier of gluten related disorders." *Nutrients* 5, no. 10 (2013): 3839–3853. https://doi.org/10.3390/nu5103839.

Bomgardner, M. M. "The problem with vanilla." *Chemical & Engineering News* 94, no. 36 (2016): 38–42. https://cen.acs.org/articles/94/i36/problem-vanilla.html.

Chapter 12

A clip of the interview with Richard Feynman is available on YouTube: https://youtu.be/ZbFM3rn4ldo.

Marazziti, D., and D. Canale. "Hormonal changes when falling in love." *Psychoneuroendocrinology* 29, no. 7 (2004): 931–936. https://doi.org/10.1016/j.psyneuen.2003.08.006.

Mercado, E., and L. C. Hibel. "I love you from the bottom of my hypothalamus: the role of stress physiology in romantic pair bond formation and maintenance." *Social and Personality Psychology Compass* 11, no. 2 (2017): e12298. https://doi.org/10.1111/spc3.12298.

Cohen, S., D. Janicki-Deverts, R. B. Turner, and W. J. Doyle. "Does hugging provide stress-buffering social support? A study of susceptibility to upper respiratory infection and illness." *Psychological Science* 26, no. 2 (2015): 135–147. https://doi.org/10.1177/0956797614559284.

Murphy, M. L., D. Janicki-Deverts, and S. Cohen. "Receiving a hug is associated with the attenuation of negative mood that occurs on days with interpersonal conflict." *PloS One* 13, no. 10 (2018): e0203522. https://doi.org/10.1371/journal.pone.0203522.

Pedersen, C. A., and A. J. Prange. "Induction of maternal behavior in virgin rats after intracerebroventricular administration of oxytocin." *Proceedings of the National Academy of Sciences of the United States of America* 76, no. 12 (1979): 6661–6665. https://doi.org/10.1073/pnas.76.12.6661.

Cho, M. M., A. C. DeVries, J. R. Williams, and C. S. Carter. "The effects of oxytocin and vasopressin on partner preferences in male and female prairie voles (Microtus ochrogaster)." *Behavioral Neuroscience* 113, no. 5 (1999): 1071–1079. https://doi.org/10.1037//0735-7044.113.5.1071.

Williams, J. R., T. R. Insel, C. R. Harbaugh, and C. S. Carter. "Oxytocin administered centrally facilitates formation of a partner preference in female prairie voles (*Microtus ochrogaster*)." *Journal of Neuroendocrinology* 6, no. 3 (1994): 247–250. https://doi.org/10.1111/j.1365-2826.1994.tb00579.x.

Baumgartner, T., M. Heinrichs, A. Vonlanthen, U. Fischbacher, and E. Fehr. "Oxytocin shapes the neural circuitry of trust and trust adaptation in humans." *Neuron* 58, no. 4 (2008): 639–650. https://doi.org/10.1016 /j.neuron.2008.04.009.

Ott, V., G. Finlayson, H. Lehnert, B. Heitmann, M. Heinrichs, J. Born, and M Hallschmid. "Oxytocin reduces reward-driven food intake in humans." *Diabetes* 62, no. 10 (2013): 3418–3425. https://doi.org/10.2337/db13-0663.

Guzmán, Y. F., N. C. Tronson, V. Jovasevic, K. Sato, A. L. Guedea, H. Mizukami, K. Nishimori, and J. Radulovic. "Fear-enhancing effects of septal oxytocin receptors." *Nature Neuroscience* 16, no. 9 (2013): 1185–1187. https://doi.org/10.1038/nn.3465.

Guzmán, Y. F., N. C. Tronson, K. Sato, I. Mesic, A. L. Guedea, K. Nishimori, and J. Radulovic. "Role of oxytocin receptors in modulation of fear by social memory." *Psychopharmacology* 231, no. 10 (2014): 2097–2105. https://doi.org/10.1007/s00213-013-3356-6.

De Dreu, C. K., L. L. Greer, G. A. Van Kleef, S. Shalvi, and M. J. Handgraaf. "Oxytocin promotes human ethnocentrism." *Proceedings of the National Academy of Sciences of the United States of America* 108, no. 4 (2011): 1262–1266. https://doi.org/10.1073/pnas.1015316108.

Guastella, A. J., S. L. Einfeld, K. M. Gray, N. J. Rinehart, B. J. Tonge, T. J. Lambert, and I. B. Hickie. "Intranasal oxytocin improves emotion recognition for youth with autism spectrum disorders." *Biological Psychiatry* 67, no. 7 (2010): 692–694. https://doi.org/10.1016/j.biopsych.2009.09.020.

Young, L. J., and C. E. Barrett. "Can oxytocin treat autism?" *Science* 347, no. 6224 (2015): 825–826. https://doi.org/10.1126/science.aaa8120.

Owen, S. F., S. N. Tuncdemir, P. L. Bader, N. N. Tirko, G. Fishell, and R. W. Tsien. "Oxytocin enhances hippocampal spike transmission by modulating fast-spiking interneurons." *Nature* 500, no. 7463 (2013): 458–462. https:// doi.org/10.1038/nature12330.

Chapter 13

Wall, T. L., H. R. Thomasson, M. A. Schuckit, and C. L. Ehlers. "Subjective feelings of alcohol intoxication in Asians with genetic variations of ALDH2 alleles." *Alcoholism: Clinical and Experimental Research* 16, no. 5 (1992): 991–995. https://doi.org/10.1111/j.1530-0277.1992.tb01907.x.

Cook, T. A., S. E. Luczak, S. H. Shea, C. L. Ehlers, L. G. Carr, and T. L. Wall. "Associations of ALDH2 and ADH1B genotypes with response to alcohol in Asian Americans." *Journal of Studies on Alcohol and Drugs* 66, no. 2 (2005): 196–204. https://doi.org/10.15288/jsa.2005.66.196.

Boffetta, P., and M. Hashibe. "Alcohol and cancer." *The Lancet Oncology* 7, no. 2 (2006): 149–156. https://doi.org/10.1016/S1470-2045(06)70577-0.

World Health Organization (WHO). *Global Status Report on Alcohol and Health* 2018. Geneva: WHO, 2018.

Bhandage, A. K. "Glutamate and GABA signalling components in the human brain and in immune cells." PhD diss., Uppsala University, 2016. *Digital Comprehensive Summaries of Uppsala Dissertations from the Faculty of Medicine* 1218: 81 pp. https://www.diva-portal.org/smash/get/diva2:918767/FULLTEXT01.pdf.

Boileau, I., J. M. Assaad, R. O. Pihl, C. Benkelfat, M. Leyton, M. Diksic, R. E. Tremblay, and A. Dagher. "Alcohol promotes dopamine release in the human nucleus accumbens." *Synapse* 49, no. 4 (2003): 226–231. https://doi.org/10.1002/syn.10226.

Cordell, B., and J. McCarthy. "A case study of gut fermentation syndrome (auto-brewery) with *Saccharomyces cerevisiae* as the causative organism." *International Journal of Clinical Medicine* 4, no. 7 (2013): 309–312. https://doi.org/10.4236/ijcm.2013.47054.

Index

Periodic Table

1 H Hydrogen								

Legend:
- 1 — Atomic Number
- H — Symbol
- Hydrogen — Name

1 H Hydrogen								
3 Li Lithium	4 Be Beryllium							
11 Na Sodium	12 Mg Magnesium							
19 K Potassium	20 Ca Calcium	21 Sc Scandium	22 Ti Titanium	23 V Vanadium	24 Cr Chromium	25 Mn Manganese	26 Fe Iron	27 Co Cobalt
37 Rb Rubidium	38 Sr Strontium	39 Y Yttrium	40 Zr Zirconium	41 Nb Niobium	42 Mo Molybdenum	43 Tc Technetium	44 Ru Ruthenium	45 Rh Rhodium
55 Cs Cesium	56 Ba Barium	57-71 Lanthanides	72 Hf Hafnium	73 Ta Tantalum	74 W Tungsten	75 Re Rhenium	76 Os Osmium	77 Ir Iridium
87 Fr Francium	88 Ra Radium	89-103 Actinides	104 Rf Rutherfordium	105 Db Dubnium	106 Sg Seaborgium	107 Bh Bohrium	108 Hs Hassium	109 Mt Meitnerium

57 La Lanthanum	58 Ce Cerium	59 Pr Praseodymium	60 Nd Neodymium	61 Pm Promethium	62 Sm Samarium	63 Eu Europium	64 Gd Gadolinium	65 Tb Terbium
89 Ac Actinium	90 Th Thorium	91 Pa Protactinium	92 U Uranium	93 Np Neptunium	94 Pu Plutonium	95 Am Americium	96 Cm Curium	97 Bk Berkelium

																	2 He Helium
												5 B Boron	6 C Carbon	7 N Nitrogen	8 O Oxygen	9 F Fluorine	10 Ne Neon
												13 Al Aluminum	14 Si Silicon	15 P Phosphorus	16 S Sulfur	17 Cl Chlorine	18 Ar Argon
28 Ni Nickel	29 Cu Copper	30 Zn Zinc	31 Ga Gallium	32 Ge Germanium	33 As Arsenic	34 Se Selenium	35 Br Bromine	36 Kr Krypton
46 Pd Palladium	47 Ag Silver	48 Cd Cadmium	49 In Indium	50 Sn Tin	51 Sb Antimony	52 Te Tellurium	53 I Iodine	54 Xe Xenon
78 Pt Platinum	79 Au Gold	80 Hg Mercury	81 Tl Thallium	82 Pb Lead	83 Bi Bismuth	84 Po Polonium	85 At Astatine	86 Rn Radon
110 Ds Darmstadtium	111 Rg Roentgenium	112 Cn Copernicium	113 Nh Nihonium	114 Fl Flerovium	115 Mc Moscovium	116 Lv Livermorium	117 Ts Tennessine	118 Og Oganesson

66 Dy Dysprosium	67 Ho Holmium	68 Er Erbium	69 Tm Thulium	70 Yb Ytterbium	71 Lu Lutetium
98 Cf Californium	99 Es Einsteinium	100 Fm Fermium	101 Md Mendelevium	102 No Nobelium	103 Lr Lawrencium